Instant Bonding Epoxy Technology

Authored by

Chunfu Chen

Henkel Technology Center - Asia Pacific
Henkel Japan Ltd., Yokohama
Japan

Instant Bonding Epoxy Technology

Author: Chunfu Chen

ISBN (Online): 978-981-5313-44-4

ISBN (Print): 978-981-5313-45-1

ISBN (Paperback): 978-981-5313-46-8

Published by Bentham Science Publishers Pte. Ltd. Singapore. All Rights Reserved.

First published in 2025.

need for a court order if at any point you breach any terms of this License Agreement. In no event will any delay or failure by Bentham Science Publishers in enforcing your compliance with this License Agreement constitute a waiver of any of its rights.

3. You acknowledge that you have read this License Agreement, and agree to be bound by its terms and conditions. To the extent that any other terms and conditions presented on any website of Bentham Science Publishers conflict with, or are inconsistent with, the terms and conditions set out in this License Agreement, you acknowledge that the terms and conditions set out in this License Agreement shall prevail.

Bentham Science Publishers Pte. Ltd.
80 Robinson Road #02-00
Singapore 068898
Singapore
Email: subscriptions@benthamscience.net

BENTHAM SCIENCE

CONTENTS

FOREWORD

Instant Bonding Epoxy Technology describes in a comprehensive manner the fundamental theory and application knowledge of epoxy adhesive formulation technology with a focus on the latest developments—for the first time —in instant bonding epoxy technology and application: UV cure cationic epoxy, dual cure hybrid epoxy, snap thermal cure epoxy, induction cure epoxy, and snap ambient cure epoxy technologies. Written by an internationally leading expert with long-time working experience in epoxy adhesive technology, Instant Bonding Epoxy Technology is an invaluable resource for researchers, formulation chemists, and application engineers related to polymer science and technology in both academia and industry.

Dr. Chen deserves praise and thanks from the adhesion and adhesive community for distilling his vast experience in this book. All those involved or interested in adhesive bonding will find this book an indispensable reference.

K.L. Mittal
Former Editor-in-Chief, Journal of Adhesion Science and Technology
Former Editor-in-Chief, Reviews of Adhesion and Adhesives
Edison, NJ, USA

PREFACE

Epoxy adhesives form very strong and durable bonds with most materials and have been widely used as typical reactive adhesives in various bonding applications. A relatively long cure time, ranging from several days at room temperature to at least tens of minutes at elevated temperature, is usually required to cure epoxy adhesives. Recently, several new types of epoxy adhesives that can bond adherends instantly at the specified curing condition while still possessing satisfactory adhesion performance after full cure have been developed and successfully used in advanced applications, such as general bonding, semiconductor packaging, and electronics assembly.

This book describes comprehensive fundamental theory and application knowledge of epoxy adhesive formulation technology. The focus is on basic chemistry, cure methods, and equipment, as well as the latest application developments in instant bonding epoxy technology. The book is divided into six chapters. The Introduction chapter covers basic chemistry in formulating epoxy adhesives with a comprehensive description of various types of epoxy resins, curing agents, and epoxy formulations. Chapter 2 describes fundamental chemistry, UV cure equipment, UV cure epoxy adhesive, and application developments in UV cure cationic epoxy technology. Chapter 3 discusses acrylate chemistry, dual cure hybrid epoxy adhesive, and application developments in UV and thermal cure hybrid epoxy technology. Chapter 4 describes the fundamental chemistry of one-component thermal cure epoxy adhesive and application developments in snap thermal cure epoxy adhesive technology. Chapter 5 discusses fundamental chemistry, the induction heating principle, and induction cure equipment, as well as the introduction to the application of induction cure epoxy technology. Chapter 6 introduces snap ambient cure epoxy technology: fast room temperature cure epoxy adhesive, cyanoacrylate hybrid epoxy adhesive, and UV and room temperature cure epoxy adhesive technology.

As the first comprehensive overview of instant bonding epoxy technology, this book is an invaluable textbook for researchers, formulating chemists, and application engineers related to polymer science and technology in both academia and industrial societies.

Chunfu Chen
Henkel Technology Center - Asia Pacific
Henkel Japan Ltd., Yokohama
Japan

CHAPTER 1

Introduction

Abstract: Epoxy adhesives are composed of epoxy resin, curing agents, and catalysts with modifiers and additives. Bisphenol A-based epoxy resin, bisphenol F-based epoxy resin, novolac type epoxy resin, aliphatic glycidyl ether epoxy resin, glycidyl amine epoxy resin, glycidyl ester epoxy resin, and cycloaliphatic epoxy resin are typical epoxy resins. Polyamine, modified polyamine, mercaptan, phenol, anhydride, tertiary amine and imidazole compounds, and cationic initiators are typical curing agents and catalysts. Epoxy adhesives are supplied in both one-component and two-component packages depending on the curing agent used and the curing method applied. Typical room temperature cure epoxy adhesives, thermal cure epoxy adhesives, UV cure epoxy adhesives, and new trends in epoxy adhesive technology developments are described.

Keywords: Curing agent, Epoxy resin, Epoxy adhesive, One-component, Room temperature cure, Two-component, Thermal cure, UV cure.

INTRODUCTION

Epoxy resins are polymer materials containing at least one carbon-oxygen-carbon three-ring known as the epoxy group, epoxide, or oxirane, whose chemical structure is shown in Fig. (**1**). In the late 1890s, epoxy resin was first discovered. In 1909, N. Prileschajew prepared an epoxy resin *via* oxidation of olefin with benzoic acid peroxide. In 1934, P. Schlack filed a patent in Germany for using amine as a curing agent for epoxy resin [1-2]. In 1936, P. Castan prepared an epoxy resin from bisphenol A and epichlorohydrin and designed a thermoset composition by using phthalic anhydride as a curing agent targeted for dental application. In the late 1940s, epoxy adhesives were commercialized in Europe and USA. Various types of epoxy resins, curing agents, and epoxy adhesives have been developed and commercialized since then. Nowadays, epoxy adhesives have been widely used as typical reactive adhesives for various bonding applications in consumer uses, general industry, construction, electronics assembly, automobile production, and aerospace and defense markets [3-25], as summarized in Table **1**.

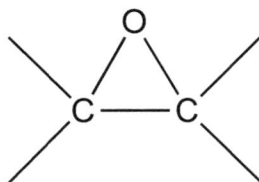

Fig. (1). Chemical structure of the epoxy group, epoxide or oxirane.

Table 1. Typical applications of epoxy adhesives.

Application	Typical Examples	Package Type	Cure Method
General industrial	Structural bonding	One-component Two-component	Room temperature cure Thermal cure UV cure
Construction	Concrete repairing Anchor bolt fixture	Two-component	Room temperature cure
Automotive industry	Structural bonding Hemming flange bonding	One-component Two-component	Thermal cure
Aerospace	Metal, Honeycomb and composite bonding, repairing	One-component Two-component	Thermal cure
Electronics	Electrically conductive Display assembly Image sensor assembly Semiconductor packaging Medical device bonding	One-component Two-component	Thermal cure UV cure Room temperature cure
Others	Sports equipment bonding Consumer applications	Two-component	Room temperature cure

Glycidyl ether epoxy resins such as Bisphenol A, Bisphenol F, novolac type, glycidyl amine epoxy resin, glycidyl ester epoxy resin, and cycloaliphatic epoxy resin are typical epoxy resins. Glycidyl-type epoxy resins are synthesized by the reaction of epichlorohydrin with phenols, alcohols, amines, or acids. Cycloaliphatic epoxy resins are prepared by the oxidation of olefins with peroxides. The preparation method and chemical structure of typical epoxy resins are summarized in Table **2**.

Table 2. Typical epoxy resins.

Type		Preparation from	Chemical Structure of the Epoxy Group
Glycidyl ether	Bisphenol A glycidyl ether	Bisphenol A + epichlorohydrin	
	Bisphenol F glycidyl ether	Bisphenol F + epichlorohydrin	
	Novalac epoxy resin	Novalac phenol + epichlorohydrin	
	Aliphatic glycidyl ether	Alcohol + epichlorohydrin	
Glycidyl amine epoxy resin		Amine + epichlorohydrin	
Glycidyl ester epoxy resin		Acid + epichlorohydrin	
Cycloaliphatic epoxy resin		Cycloolefin + peroxide	

Bisphenol A-Based Epoxy Resins

Bisphenol A-based epoxy resin, also called glycidyl ether of bisphenol A and often abbreviated as BADGE or DGEBA, was the first commercialized epoxy resin. It is still the most standard and widely used epoxy resin, constituting the majority, estimated over 75% in sales volume, of all epoxy resins used today. Bisphenol A epoxy resin is typically prepared by the reaction of bisphenol A and epichlorohydrin at 70 – 80°C under alkaline conditions, as illustrated in Fig. (**2**) [26]. Chemical structure and key features of functional groups of bisphenol A epoxy resin are illustrated in Fig. (**3**) [27], indicating potentially balanced properties of good reactivity, good chemical resistance, and high thermal resistance, as well as high adhesion.

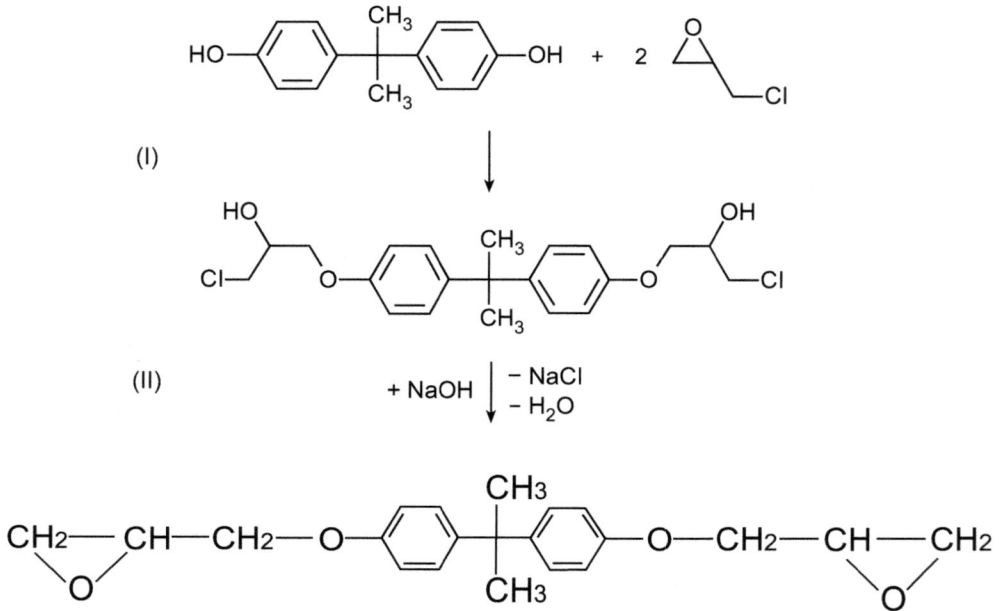

Fig. (**2**). Synthesis of DGEBA from bisphenol A and epichlorohydrin.

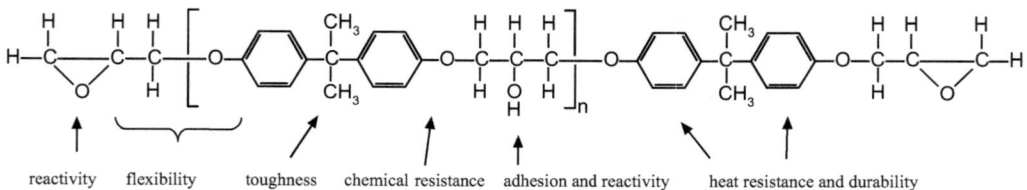

Fig. (**3**). Chemical structure and key features of DGEBA.

Different grades of bisphenol A epoxy resins in either liquid or solid state with various viscosity or melting points, EEW (abbreviated for epoxy equivalent weight), and purity have been commercialized and supplied by major epoxy resin manufacturers. Typical unmodified bisphenol A glycidyl ether epoxy resin products are listed in Table **3**. Epoxy resin with an EEW of 182 to 192 is the market standard product, most widely used for general-purpose applications.

Table 3. Typical commercial bisphenol A-based epoxy resin.

Type	n value[1]	EEW, g/eq	Viscosity, mPas/25°C	Softening Point, °C	Commercial Products[2]
Distilled	<0.1	170-175	4000-6000	-	D.E.R. 332, Araldite MY 790-1, EPON 825, EPICLON EXA-850CRP
Liquid	~0.1	176-186	6500-9500	-	EPON 826, YD-126
Liquid	~0.2	180-190	8000-11000	-	EPON 827, jER 827
Standard liquid	~0.2	182-192	11000-14000	-	D.E.R. 331, Araldite GY 6010, EPON 828, jER 828, EPICLON 850, YD-828
Liquid	~0.3	205-225	19000-24000	-	D.E.R. 317, EPON 830, YD-128S
Semi-solid	~0.5	230-250	Semi-solid	-	D.E.R. 337, EPON 834, EPICLON 860
Solid	~2.0	450-500	-	60-70	D.E.R. 661, EPON 1001F, jER 1001, YD-011
Solid	~3.5	600-700		75-85	D.E.R. 662, Araldite GT 7072, EPON 1002F, jER 1002, YD-012

*1 n value is the average reputing unit number, as shown in Fig. (**3**).

*2 D.E.R. products are supplied by Olin Corporation, Araldite products are supplied by Huntsman Corporation, EPON products are supplied by Westlake Chemical Corporation, EPICLON products are supplied by DIC Corporation, jER products are supplied by Mitsubishi Chemical Corporation, YD- products are supplied by Kukdo Chemical Co., Ltd.

EEW is the weight of epoxy resin per one epoxy group, as shown in Eq. **1.1**. EEW is a very important parameter in epoxy technology. It is measured by titration according to standard methods ASTM D 1652 or EN ISO 3001.

EEW = Molar mass of epoxy resin / Number of epoxy group

Equation 1.1. EEW calculation.

The EEW of the epoxy resin mixture can be calculated based on the EEW value of each epoxy resin component formulated, as illustrated in Eq. **1.2** for mixture examples of epoxy resin a, epoxy b, and epoxy c. The total weight is the sum of the weight of all epoxy resins and other components, such as fillers that do not have an epoxy group.

EEW of epoxy mixture = Total weight / (Weight$_a$ /EEW$_a$ + Weight$_b$ /EEW$_b$ + Weight$_c$ /EEW$_c$)

Equation 1.2. EEW calculation of epoxy mixture.

Chlorine content is a very important parameter used in epoxy technology, especially in electronics applications. All epoxy resins prepared from epichlorohydrin will inevitably contain chlorine. Chlorines, especially the hydrolyzable ones, will potentially damage electronic properties in actual use at strict high temperature and high humidity conditions. Additionally, due to the concern for health and environmental issues caused by halogens, the International Electrochemical Commission (IEC) set forth a standard called IEC 61249-2-21, which defines halogen-free as the following: chlorine ≤ 900 ppm, bromine ≤ 900 ppm, and total halogens ≤ 1500 ppm [28]. Halogen-free grade epoxy adhesives are increasingly required, especially in the semiconductor packaging and electronics assembly applications. Preparation of low chlorine content grade epoxy resins will require an additional purification process or specially designed synthesis method, and thus, low chlorine content epoxy resin products, normally called electronics grade products, have much higher cost. Table **4** compares chlorine contents of different grade bisphenol A-based epoxy resins supplied by Mitsubishi Chemical Corporation.

Bisphenol F-Based Epoxy Resins

Bisphenol F-based epoxy resin, also called bisphenol F glycidyl ether and abbreviated as DGEBF, is synthesized by the reaction of bisphenol F and epichlorohydrin. Fig. (**4**) shows the typical chemical structure of bisphenol F-based epoxy resin. Bisphenol F epoxy resin has lower viscosity and better solvent and chemical resistance than standard bisphenol A epoxy resin. Typical unmodified bisphenol F glycidyl ether epoxy resin products are listed in Table **5**.

Table 4. Chlorine content comparison of different grade bisphenol A-based epoxy resins.

Type	EEW	Viscosity mPas/25°C	Chlorine Content, ppm	
			Hydrolyzable	Total
jER 828	184-194	12000-15000	1200	3000
jER 828EL	184-194	12000-15000	-	1700
jER 828US	184-194	12000-15000	700	1500
jER 825	170-180	4000-7000	700	1700
YL 980	180-190	12000-15000	150	300

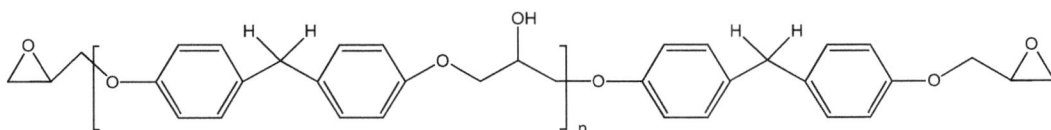

Fig. (4). Chemical structure of bisphenol F epoxy resin.

Table 5. Typical commercial bisphenol F-based epoxy resin.

Type	EEW, g/eq	Viscosity, mPa.s/25°C	Softening Point, °C	Commercial Products*
Low Cl	154-170	1400-3200	-	YDF-870GS
Distilled	155-165	1000-3000	-	YDF-8170C, Araldite PY 306, EPICLON EXA-830LVP
Standard liquid	160-180	3000-4000	-	D.E.R. 354, Araldite GY 282, EPON 862, EPICLON 830, YDF-170

(Table 5) cont.....

Low crystallization	160-180	3000-4500		EPICLON 835, YDF-170N
Bis A/F blend	170-180	5000-7000	-	D.E.R. 351, Araldite GY-281,
Distilled Bis A/F blend	160-170	1900-2600		ZX-1059
Solid	950-1200	-	87	jER 4005P

* D.E.R. products are supplied by Olin Corporation, Araldite products are supplied by Huntsman Corporation, EPON products are supplied by Westlake Chemical Corporation, EPICLON products are supplied by DIC Corporation, jER products are supplied by Mitsubishi Chemical Corporation, YDF- and ZX- products are supplied by Nippon Steel Chemical & Material Co., Ltd.

Novolac Epoxy Resins

Novolac epoxy resins, including cresol and phenol novolac epoxy resin, also called novolac glycidyl ether and sometimes abbreviated as NOGE, are synthesized by the reaction of epichlorohydrin with novolac phenols. The typical chemical structure of novolac epoxy resin, including cresol novolac epoxy resin and phenol novolac epoxy resin, is shown in Fig. (**5**). Novolac epoxy resin has multi-functional epoxy groups and thus possesses high thermal resistance properties. Novolac epoxy resin is usually used in combination with bisphenol A or bisphenol F epoxy resin for easy handling in actual uses. Typical cresol and phenol novolac epoxy resin products are listed in Table **6**.

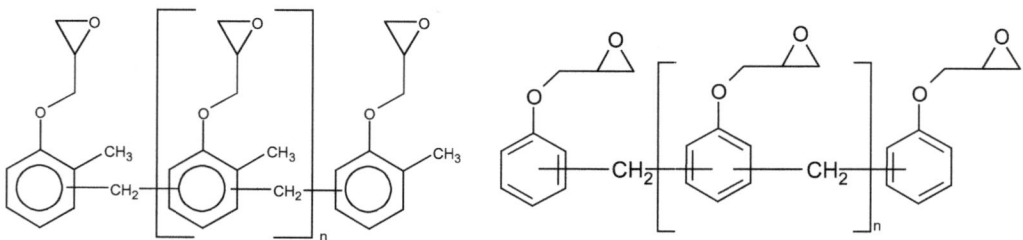

Fig. (5). Chemical structures of cresol (left) and phenol (right) novolac epoxy resins.

Table 6. Typical commercial cresol and phenol novolac epoxy resin.

Type	Functionality	EEW, g/eq	Viscosity, mPa.s/25°C	Softening Point, °C	Commercial Products*
Phenol novalac	2.4	165-175	5700-6800	-	D.E.N. 424
	2.5	169-175	9500-12500	-	D.E.N. 425
	2.6	167-177	18000-23000	-	D.E.N. 426
	2.7	168-178	20000-26000	-	Araldite EPN 9850
	2.8	172-179	1100-1700 mPa.s/52°C	-	D.E.N. 431, Araldite EPN 1179, jER 152
	3.6	176-181	31000-40000 mPa.s/52°C	-	D.E.N. 438, Araldite EPN 1180, jER 154
	3.9	191-210	15000-35000 m.Pas/71°C	-	D.E.N. 439
	4.4	186-192	20000-45000 m.Pas/71°C	-	D.E.N. 440
	-	170-190	Semi-solid	-	EPON 154, YDPN 638, EPICLON N-740
Cresol novalac	-	200-215	-	60-68	EPICLON N-660
	-	200-215	-	64-72	EPICLON N-665
	-	200-215	-	64-72	EPICLON N-665EXP
	5.1	190-220	-	75-85	EPICLON N-685, EPON 164, Araldite ECN 1280, YDCN 700-10

* Araldite products are supplied by Huntsman Corporation, EPON products are supplied by Westlake Chemical Corporation, EPICLON products are supplied by DIC Corporation, jER products are supplied by Mitsubishi Chemical Corporation, YDPN- and YDCN- products are supplied by Nippon Steel Chemical & Material Co., Ltd.

Glycidyl Amine-Based Epoxy Resins

Typical glycidyl amine-based epoxy resins commercialized are di-functional glycidyl amine, tri-functional glycidyl amine, tetra-functional glycidyl amine, and tetra-functional aliphatic amine-based epoxy resins. They are usually prepared by the reaction of epichlorohydrin with amines. Their chemical structures are shown in Fig. (**6**). Di-functional and tri-functional glycidyl amine epoxy resins have relatively low viscosity and are often used as reactive diluents, especially for high temperature-resistant applications. Tetra-functional glycidyl amine epoxy resin possesses extremely high glass transition temperature and excellent mechanical properties after a full cure. Tetra-functional amine-based epoxy resins are commonly used as base resins for applications that require high performance. Tetra-functional aliphatic amine-based epoxy resin has lower viscosity than aromatic amine-based epoxy resins. Typical glycidyl ether and aliphatic amine epoxy resin products are listed in Table **7**.

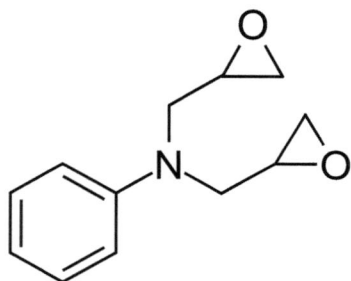

Diglycidyl aniline

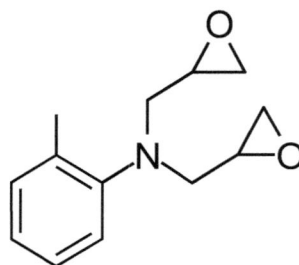

2-Methyl diglycidyl aniline

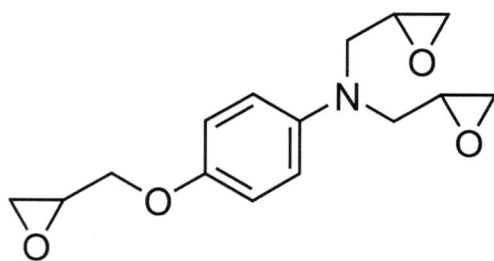

N, N ,O -Triglycidyl-p -aminophenol

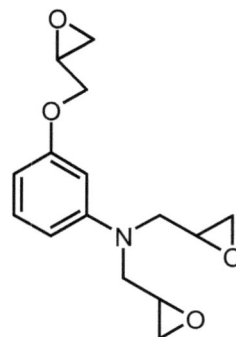

N,N,O-Triglycidyl of m-amino phenol

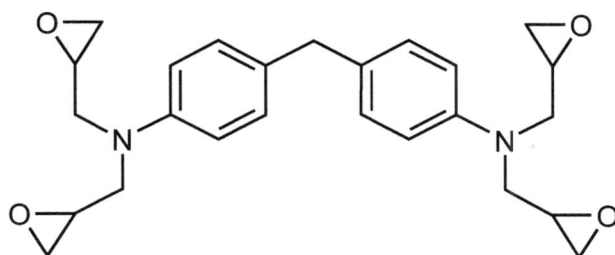

Tetraglycidyl diamino diphenylmethane

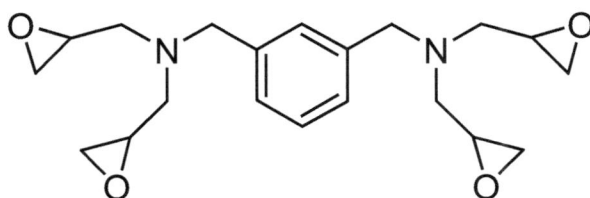

N,N',N,N'-Tetraglycidyl-m-xylenediamine

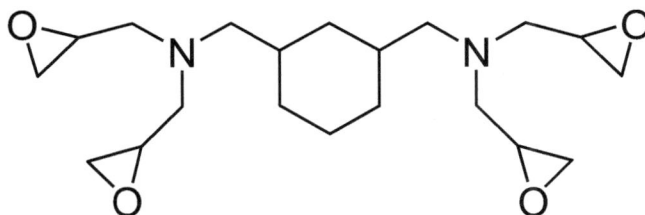

N,N,N',N'-tetrakis(2,3-epoxypropyl)cyclohexane-1,3-dimethylamine

Fig. (6). Chemical structures of typical glycidyl amine epoxy resins.

Table 7. Typical commercial glycidyl amine-based epoxy resins.

Type	Chemical Name	CAS No.	EEW, g/eq	Viscosity, mPa.s/25°C	Commercial Products*
di-functional	N,N-Diglycidyl-o-toluidine	40027-50-7	115-135	30-80	GOT

(Table 7) cont.....

	Diglycidyl aniline	2095-06-9	125-145	100-160	GAN
Tri-functional	N ,N ,O -Triglycidyl-p -aminophenol	5026-74-4	95-105	550-850	Araldite MY 0501, jER 630, Adeka Resin EP-3950S
			105-115	2000-5000	Araldite MY 0510,
	N,N,O-Triglycidyl of m-amino phenol	40027-50-7	102-110	7000-13000	Araldite MY 0600
			95-102	1000-6000	Araldite MY 0610
Tetra-functional aromatic	Tetraglycidyl diamino diphenylmethane	28768-32-3	111-117	3000-6000 mPa.s/50 °C	Araldite MY 721
			117-134	7000-19000 mPa.s/50 °C	Araldite MY 720, jER 604
Tetra-functional aliphatic	N,N,N',N'-tetrakis(2,3-epoxypropyl)cyclohexane-1,3-dimethylamine	65992-66-7	95-110	1600-3000	Tetrad-C
	N,N',N,N'-Tetraglycidyl-m-xylenediamine	63738-22-7	95-110	1600-3000	Tetrad-X, Erisys GA-240

* Araldite and Erisys products are supplied by Huntsman Corporation, jER products are supplied by Mitsubishi Chemical Corporation, GOT and GAN are supplied by

Nippon Kayaku Co., Ltd., Tetrad products are supplied by Mitsubishi Gas Chemical Co., Ltd., Adeka Resin products are supplied by Adeka Corporation.

Mono-functional Aromatic and Aliphatic Glycidyl Ether Epoxy Resins

Mono-functional aromatic glycidyl ether epoxy resins are prepared by the reaction of epichlorohydrin with phenols. Aliphatic glycidyl ether epoxy resins are prepared by the reaction of epichlorohydrin with alcohol. Mono-functional aromatic and aliphatic glycidyl ether epoxy resins usually have low viscosity and are used mainly as reactive diluents to lower viscosity for better handling property adjustment. The chemical structure of typical glycidyl ether epoxy resins synthesized from alcohols and epichlorohydrin is illustrated in Fig. (**7**). The chemical structure of typical glycidyl ether epoxy resins synthesized from phenols and epichlorohydrin is illustrated in Fig. (**8**). Their chemical name, CAS No., EEW value, viscosity, and typical commercial products are summarized in Table **8**.

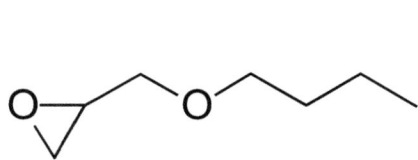

Butyl glycidyl ether (BGE)

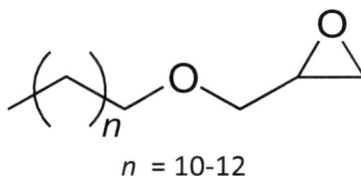

n = 10-12

Aliphatic glycidyl ether (C12-C14)

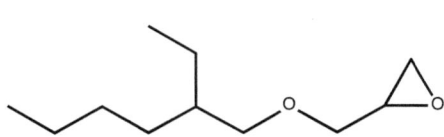

2-ethylhexyl glycidyl ether

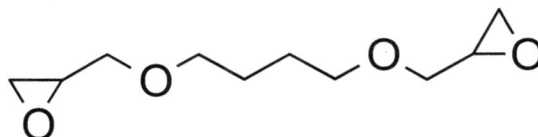

1,4 - Butanediol diglycidyl ether

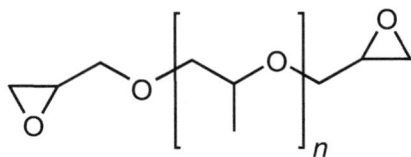

Polypropylene glycol diglycidyl ether

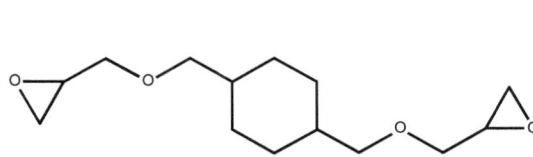

1,4-Cyclohexane dimethanol diglycidyl ether

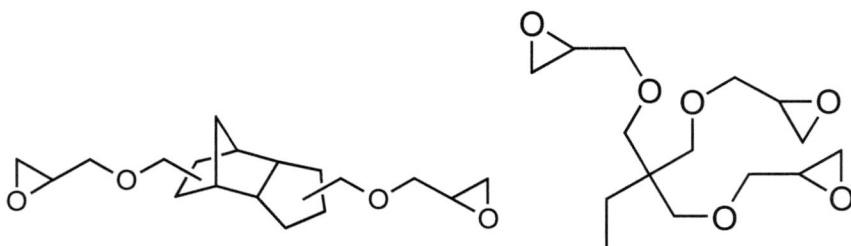

Dicyclopentadiene dimethanol diglycidyl ether   Trimethylolpropane triglycidyl
                                                                          ether

Fig. (7). Chemical structure of typical aliphatic glycidyl ether.

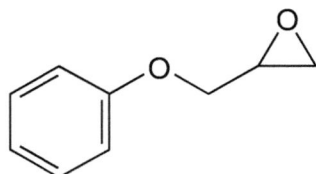

Phenyl glycidyl ether   Cresol glycidyl ether

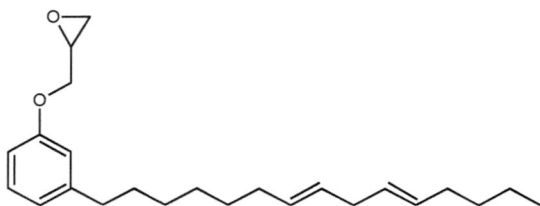

3-alkyl phenol glycidyl ether   p-tertiary butyl phenyl glycidyl ether

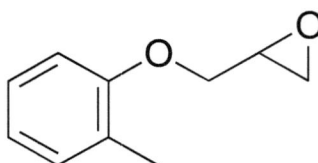

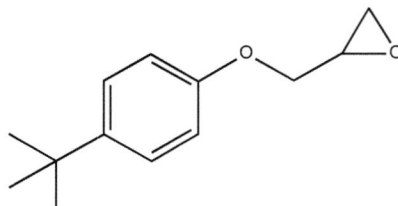

Fig. (8). Chemical structure of mono-functional aromatic glycidyl ether.

Table 8. Typical commercial epoxy reactive diluents.

Type	Chemical Name	CAS No.	EEW, g/eq	Viscosity, mPa.s/25°C	Commercial Products*
	n-butyl glycidyl ether (BGE)	2426-08-6	145-155	1-5	Heloxy 61, Epodil 741

(Table 8) cont.....

Aliphatic glycidyl ether	Aliphatic glycidyl ether (C12-C14)	68609-97-2	275-300	5-10	Heloxy 8, Araldite DY-E, Epodil 748
	2-ethylhexyl glycidyl ether	2461-15-6	205-235	1-5	Heloxy 116, Araldite DY-A, Epodil 746
	Cyclohexane dimethanol diglycidyl ether	14228-73-0	60-90	167-179	Araldite DY-C
	1,4 butanediol diglycidyl ether	2425-79-8	117-125	15-25	Araldite DY-D, Epodil 750
	Polypropylene glycol diglycidyl ether (200)	26142-30-3	185-215	60-75	D.E.R. 736
	Polypropolene glycol diglycidyl ether (400)		300-330	40-75	D.E.R. 732
	1,4-Cyclohexane dimethanol diglycidyl ether	14228-73-0	145-165	45-75	Erisys GE-22
	Dicyclopentadiene dimethanol diglycidyl ether	50985-55-2	170	230	EP-4088S
	Trimethylolpropane triglycidyl ether	30499-70-8	111-143	100-200	Araldite DY-T
Mono-functional aromatic glycidyl ether	Phenyl glycidyl ether (PGE)	122-60-1	150-165	4-8	Heloxy 63
	o-Cresyl glycidyl ether	2210-79-9	170-195	5-10	Heloxy 62, Araldite DY-K, Epodil 742
	3-alkyl phenol glycidyl ether	68413-24-1	~ 490	50	Heloxy 64, Araldite DY CNO, NC-513

(Table 8) cont.....

	p-tertiary butyl phenyl glycidyl ether	3101-60-8	210-240	18-28		Heloxy 65, Araldite DY-P

* Heloxy products are supplied by Westlake Chemical Corporation, Araldite and Erisys products are supplied by Huntsman Corporation, D.E.R. products are supplied by Olin Corporation, NC-513 is supplied by Cardolite Corporation, Epodil products are supplied by Evonik Corporation, EP-4088S is supplied by ADEKA Corporation.

Glycidyl Ester-Based Epoxy Resins

Glycidyl ester-based epoxy resins are synthesized by the reaction of epichlorohydrin with carboxylic acids, as illustrated in Fig. (9) [29]. Anhydride is often selected as the curing agent for glycidyl ester-based epoxy resin. Its cured epoxy resin shows good insulation, high thermal resistance, and good UV-resistant performance. Fig. (10) shows the chemical structure and key features of typical glycidyl ester-based epoxy resins.

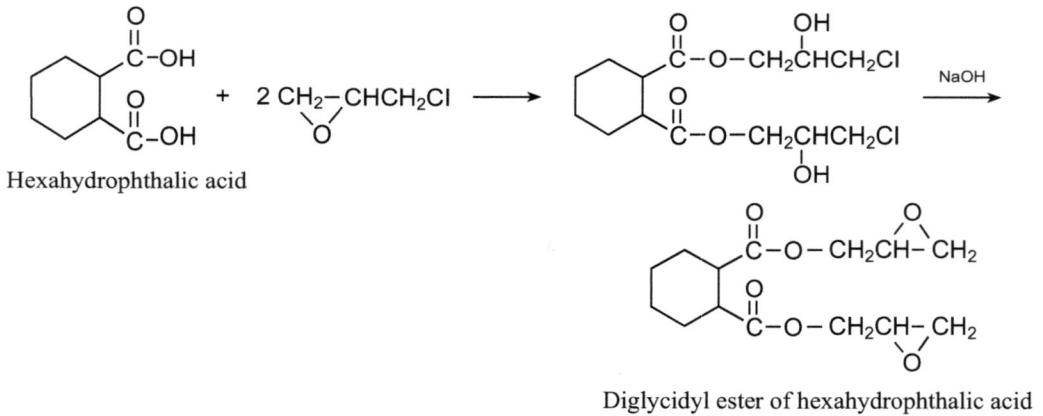

Hexahydrophthalic acid

Diglycidyl ester of hexahydrophthalic acid

Fig. (9). Synthesis of glycidyl ester epoxy resin from carboxylic acid and epichlorohydrin.

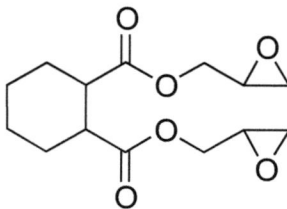

Viscosity: 700-1000 mPas/25°C
CAS No.: 5493-45-8
EEW: 144-157
Standard

Diglycidyl 1,2-cyclohexanedicarboxylate

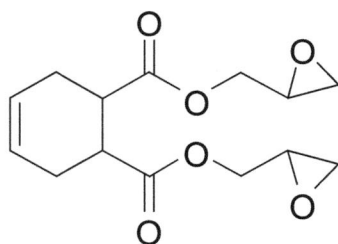

Viscosity: 500 mPas/25°C
CAS No.: 21544-03-6
EEW: 150-175
Good adhesion

Diglycidyl 4-cyclohexene-1,2-dicarboxylate

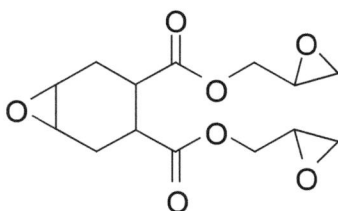

Viscosity: 10000 mPas/25°C
CAS No.: 25293-64-5
EEW: 110-135
High thermal resistance

4,5-Epoxycyclohexane-1,2-dicarboxylic acid diglycidyl ester

Fig. (10). Chemical structure and key features of typical glycidyl ester type epoxy resins.

Cycloaliphatic Epoxy Resins

Cycloaliphatic epoxy resin is normally synthesized by the oxidation of cycloolefin with peroxide, as illustrated in Fig. (**11**) [30]. As can be seen, there is no chlorine substance involved in the preparation process, so cycloaliphatic epoxy resin is completely halogen-free epoxy resin. Anhydrides or cationic initiators are the main curing agents or initiators. Cured resins offer good weathering and thermal-resistant performance. Physical properties of cured cycloaliphatic epoxy resin compared with bisphenol A epoxy resin cured with anhydride are shown in Table **9** [31].

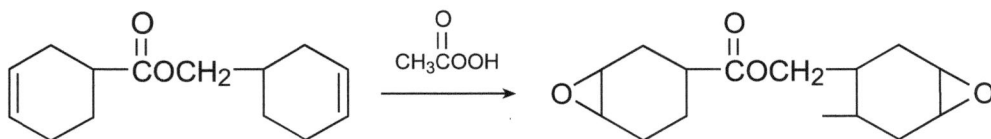

Fig. (11). Synthesis of cycloaliphatic epoxy resin by oxidation of olefin.

Table 9. Physical properties of cured cycloaliphatic and bisphenol A epoxy compositions.

Composition	1	2	3	4	5
EHPE-3150[1]	100				
jER 1001[2]	93	100	100	100	100
jER 828[3]	0.97	36	89	158	123
Arldite PT810[4]		0.68	0.95	1.29	1.12
Celloxide 2021[5]					
MH-700[6]					
BDMA					
Tg^{7}, °C	224	115	142	200	196
Water absorption, %	2.93	3.06	2.05	5.49	4.42
Refractive index	1.59	1.62	1.6	1.54	1.57
Transmission rate, %	85	78	57	79	80

*1. Multifunctional cycloaliphatic epoxy resin supplied by Daicel Corporation.

2. Solid bisphenol A epoxy resin (EEW=500), supplied by Mitsubishi Chemical Corporation.

3. Standard bisphenol A epoxy resin (EEW=190), supplied by Mitsubishi Chemical Corporation.

4. Triglycidyl isocyanurate type epoxy resin, supplied by Huntsman Corporation.

5. (3',4'-Epoxycyclohexane)methyl-3,4-epoxycyclohexylcarboxylate, supplied by Daicel Corporation.

6. Methyl hexahydrophthalic anhydride, supplied by New Japan Chemical Co., Ltd.

7. TMA method. Cure condition: 120°C x 1 hour + 240°C x 1 hour.

Flexible Epoxy Resins

Cured epoxy resin is a highly cross-linked thermoset polymer. It is rigid and strong but relatively brittle and not tough enough, thus tending to crack easily in actual uses. Flexible epoxy resins are prepared by chemically introducing a soft structure into epoxy resins to improve the toughness of the epoxy structure [32-35]. Isocyanate-modified, polyurethane-modified, polyol-modified, CTBN (carboxyl-terminated butadiene nitrile)-modified, and dimeric acid-modified epoxy resins are typical flexible epoxy resins supplied in the market [36-39]. Fig. (**12**) illustrates their chemical structure. However, the glass transition temperature of most flexible epoxy resins becomes lower due to the introduction of flexible chemical structures, often resulting in decreased thermal resistance. Gel time, physical properties, and adhesion strength of polyurethane-modified epoxy resin cured with a modified amine curing agent are compared with those of standard bisphenol A epoxy resin in Table **10** [40]. Tensile strength was lower, and elongation became much longer, indicating a much more flexible structure than bisphenol A epoxy resin.

Isocyanate-modified epoxy resin

Polyurethane-modified epoxy resin

Bisphenol A bis (propyleneglycolglycidylether)ether

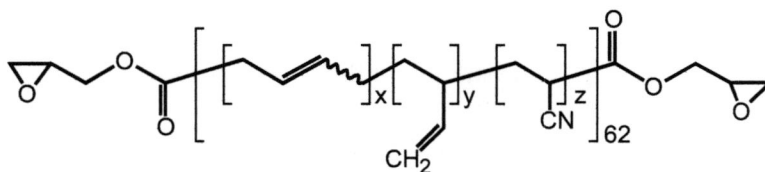

CTBN-modified epoxy resin

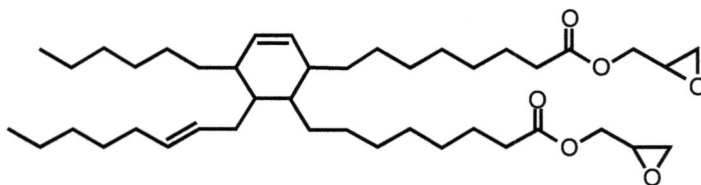

Dimeric acid-modified epoxy resin

Fig. (12). Chemical structure of typical flexible epoxy resins.

Table 10. Property comparison of polyurethane modified epoxy resin with bisphenol A epoxy compositions cured with modified amine.

Composition	1	2	3
EPU-6[1]	100		
EPU-11[2]	35	100	100
DGEBA (EEW=190)		35	40
EH-220[3]			
Gel time, min @25°C	40	60	17
Tensile strength, MPa	8.8	3.9	48.7
Elongation at break, %	32	132	1.7
Lap shear strength, MPa	15.5	9.6	9.3
Water absorption, %	1.2	0.7	0.56

1. Polyurethane-modified epoxy resin, supplied by ADEKA Corporation.

2. Polyurethane-modified epoxy resin, supplied by ADEKA Corporation.

3. Modified aliphatic amine curing agent supplied by ADEKA Corporation.

Specialty Epoxy Resins

Several other specialty epoxy resins have been developed and commercialized. Fig. (**13**) shows the chemical structure of these specialty epoxy resins. DCPD (dicyclopentadiene) novolac epoxy resins are solid-type epoxy resins. They are hydrophobic structures and thus have better humidity resistance performance. Naphthalene-type epoxy resins have a rigid, strong structure that possesses very good thermal resistance properties with very high Tg. Bi-functional naphthalene epoxy resin HP 4032 supplied by DIC Corporation is a viscous liquid epoxy resin with a viscosity of 107000 mPas/25°C and an EEW of 152 g/eq. It has higher Tg and can cure faster than bisphenol A epoxy resin. Biphenol-based epoxy resin is a crystalline solid at room temperature with very low melting viscosity at elevated temperatures. Biphenol-based epoxy resin is mainly used for molding compounds in semiconductor packaging because of its high filler loading capability. Triglycidyl isocyanurate type epoxy resin is also a solid epoxy resin with very good weather resistance ability, thus, it is mainly used for outdoor powder coating applications.

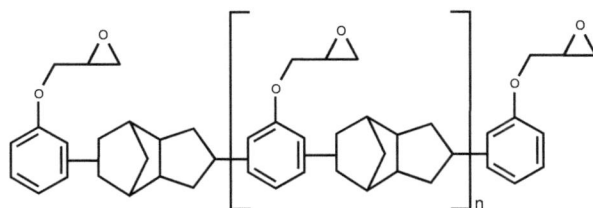

DCPD (Dicyclopendiene) novolac epoxy resin

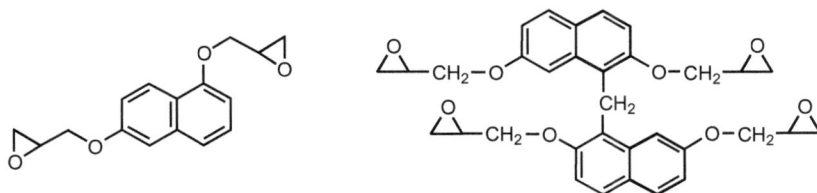

Bi-functional (left) and tetra- (right) functional naphthalene-type epoxy resin

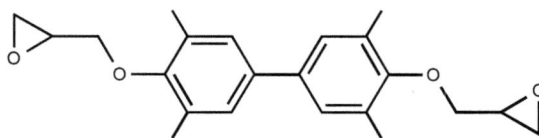

Glycidyl ether of tetramethyl biphenol

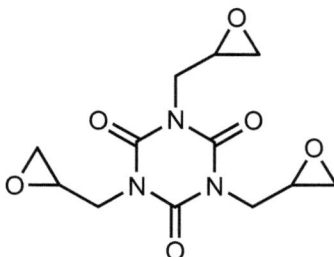

Triglycidyl isocyanurate-type epoxy resin

Fig. (13). Chemical structure of typical specialty epoxy resins.

CURING AGENTS AND CATALYSTS

The epoxide group of epoxy resins is chemically very active. Epoxy resin can react, almost equivalently, with active hydrogen in polyamines, mercaptan compounds, phenols, and anhydrates *via* a polyaddition mechanism. Epoxy resin can polymerize homogeneously *via* anionic polymerization mechanism by initiating it with Lewis bases such as tertiary amines or imidazole compounds. It can also polymerize *via* cationic polymerization by initiating it with Lewis acids such as onium salts and iodonium salts. Table **11** lists typical curing agents, catalysts, and initiators used for curing epoxy resins.

Table 11. Typical epoxy resin curing agents, catalysts, and initiators.

Polymerization Mechanism	CURING AGENT, CATALYST
Polyaddition	Polyamines
	Modified polyamines
	Mercaptans

		Phenols
		Anhydrides
Anionic		Tertiary amines
		Imidazole compounds
Cationic		Amine-BF$_3$ complex
		Onium salts, Iodonium salts

Polyamines

Polyamines, including aliphatic, cycloaliphatic, and aromatic types, are the most commonly used curing agents for curing epoxy resin. Their active hydrogen can react with the epoxy group almost stoichiometrically at certain suitable conditions, depending on the chemical structure of polyamines. As shown in Fig. (**14**) [41], primary amine will at first react with the epoxy group to form a secondary amine and a secondary alcohol. The secondary amine will further react with the epoxy group to form a tertiary amine and two secondary alcohols. Aliphatic polyamine can react with the epoxy group rapidly at room temperature. Cycloaliphatic amines can also react with epoxy resins at room temperature, but post-thermal cure is usually required for full cure achievement. Aromatic amines will need thermal curing at relatively high temperatures to cure epoxy resin.

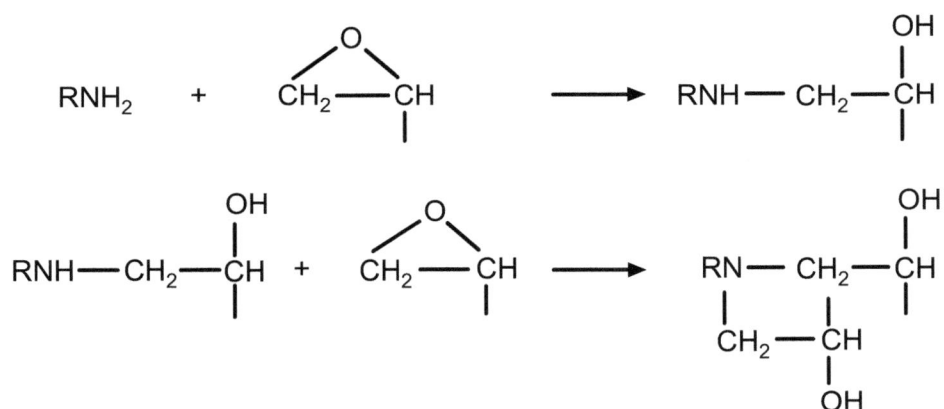

Fig. (14). Polyaddition mechanism of epoxy and polyamine.

AHEW is active hydrogen equivalent weight, representing weight in grams of amine curing agent per one active amine hydrogen. AHEW is a very important parameter in epoxy resin technology. The calculation of AHEW is shown in Eq. **1.3**.

AHEW = Molar mass of amine curing agent/Number of active amine hydrogen

Equation 1.3. EEW calculation.

AHEW of amine curing agent mixture can be calculated based on the AHEW value of each amine curing agent formulated, as illustrated in Eq. **1.4** for mixture example of amine curing agent a, b, and c. Total weight is the sum of the weight of all amine curing agent mixtures, including other components such as fillers that do not have an epoxy group.

AHEW of amine mixture = Total weight / (Weight$_a$ /AHEW$_a$ + Weight$_b$ /AHEW$_b$ + Weight$_c$ /AHEW$_c$)

Equation 1.4. AHEW calculation of amine curing agent mixture.

Aliphatic polyamines are typically used as room-temperature curable curing agents for epoxy resin. The chemical structure of typical aliphatic polyamine curing agents is shown in Fig. (**15**). Their viscosity, AHEW, cure behavior, and heat deflection temperature (HDT) are summarized in Table **12**. They are low-viscosity liquids with low AHEW values. Their reactivity is high, and the pot life at room temperature is less than 60 minutes at 25°C. Cured epoxy resin of ethylenediamines such as DETA, TETA and TEPA is very rigid with relatively high Tg. Cured epoxy resin of polyether amines becomes much more flexible with lower Tg.

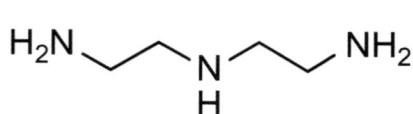

DETA (diethylene triamine)   TETA (trieethylene tetramine)

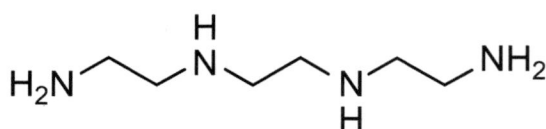

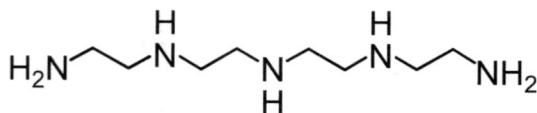

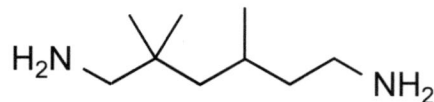

TEPA (tetraethylene pentaamine)   TMD (trimethylhexamethylenediamine)

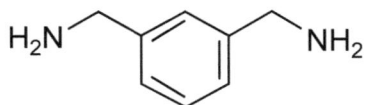

MXDA (m-xylene diamine)

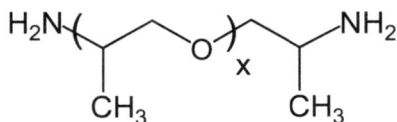

Polyether amine

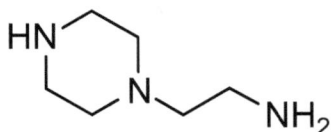

N-AEP (N-aminoethylpiperazine)

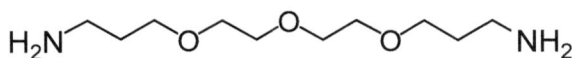

4,7,10-Trioxatridecane-1,13-diamine

Fig. (15). Chemical structures of typical aliphatic polyamine curing agents.

Table 12. Typical aliphatic polyamine-type curing agents.

-	Viscosity mPa.s/25 °C	AHEW g/eqNH	Pot-life min*	Cure condition, at room temperature	HDT °C	Manufacturer
DETA	5.6	20.7	20	4 days	115	BASF, Huntsman, Tosoh
TETA	19.4	24.4	20-30	4 days	115	BASF, Huntsman, Tosoh
TEPA	51.9	27.1	30-40	7 days	115	Huntsman, Tosoh
TMD	5.6	39.6	30-40	-	-	Evonik
MXDA	6.8	34.1	20-30	7 days	115	Mitsubishi Gas Chemical

(Table 12) cont.....

N-AEP		43	20-30	3 days	103	Huntsman, Tosoh
4,7,10-Trioxatri decane-1,13-diamine	10	55	-	-		BASF, Evonik

*Standard DGEBA (EEW=190) based, 100g scale.

Pot-life is defined as the amount of time it takes for an initial mixed viscosity of epoxy product to reach twice the initial viscosity. The timing starts from the moment the product is mixed and is measured normally at room temperature. The pot life is an important parameter, representing one key handling property of the curing agent and epoxy product.

HDT, the abbreviation for heat deflection temperature, is the temperature at which the epoxy material deforms under a constant testing load. HDT is used as a temperature resistance parameter.

Cycloaliphatic amines can react with epoxy resins at room temperature but need post thermal cure for full cure achievement. Cured epoxy resin with cycloaliphatic amines possesses higher glass transition temperature than aliphatic amine curing agents and is suitable for high-temperature resistance required applications. Fig. (**16**) shows the chemical structures of typical cycloaliphatic polyamines, IPDA (isophorone diamine), NBDA (bis(aminomethyl)norbornane), PACM (4, 4'-diaminodicyclohexyl methane), and DMDC (dimethyldicycan). Their viscosity, AHEW value, pot life, cure condition, HDT value, and major suppliers are summarized in Table **13**.

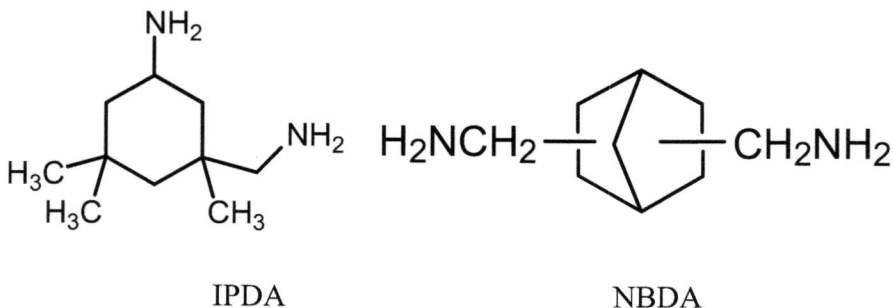

IPDA   NBDA

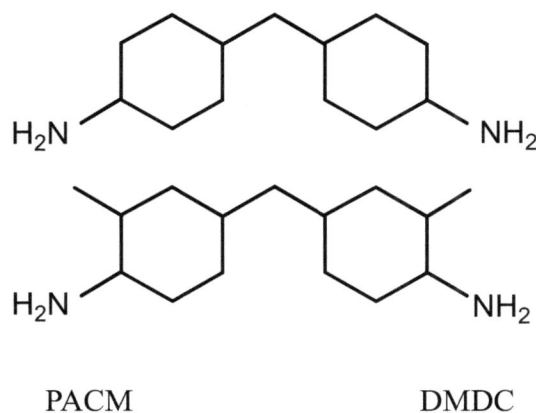

PACM   DMDC

Fig. (16). Chemical structures of typical cycloaliphatic amine curing agents.

Table 13. Typical cycloaliphatic amine-type curing agents.

-	Viscosity mPa.s/25°C	AHEW g/eqNH	Pot-life min*	Cure Condition	HDT °C	Manufacturer
IPDA	18	42.6	60	1 h @150°C	149	BASF, Evonik.
NBDA	20	38.6	20-30	2 h @120°C	130	Mitsui Chemical.
PACM	-	52.5	30-40		115	BASF, Evonik.
DMDC	60	59.5	-	2 h @150°C	150	BASF.

*Standard DGEBA (EEW=190) based, 100g scale.

Due to conjugation structure, aromatic amines have lower electron density and are, therefore, much less reactive than aliphatic and cycloaliphatic amines with epoxy resins. They have much longer pot life but will need a thermal cure at relatively high temperatures. Cured epoxy resin with aromatic amines usually possesses very good thermal resistance and mechanical properties and is suitable for use in applications that have high performance, such as aerospace assembly. Fig. (17) shows the chemical structures of typical aromatic amine curing agents. Their viscosity, AHEW value, pot life, cure condition, HDT value, and major suppliers are summarized in Table **14**.

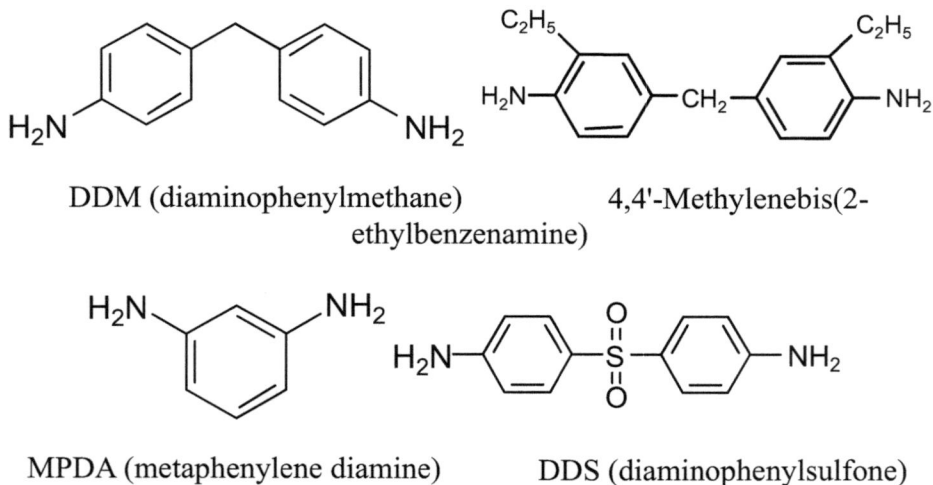

DDM (diaminophenylmethane)   4,4'-Methylenebis(2-
ethylbenzenamine)

MPDA (metaphenylene diamine)   DDS (diaminophenylsulfone)

Fig. (17). Chemical structure of typical aromatic amine curing agents.

Table 14. Typical aromatic amine-type curing agents.

-	Viscosity mPa.s/25°C	AHEW g/eqNH	Pot-life min*	Cure condition	HDT °C	Manufacturer
DDM	Solid	49.6	8	150°C 4hrs	150	BASF, Evonik
4,4'-Methylenebis(2-ethylbenzenamine)	2000-3000	-	-	-	-	Nippon Kayaku
MPDA	Solid	31	6	150°C 4hrs	150	1
DDS	Solid	62	1 year	200°C 4hrs	190	BASF, Evonik

Modified Polyamines

Modified polyamines are prepared by chemically modifying polyamines for safety, handling, and performance property improvement. Table **15** lists typical modified polyamines, their synthesis methods, and their advantages.

Table 15. Typical modified polyamine-type curing agents.

Modified Polyamine	Synthesis Mechanism	Advantages
Polyamide	Condensation reaction of polyamine with dimer acid.	Low toxicity Long pot-life Good adhesion and flexibility
Amidoamine	Condensation reaction of polyamine with monobasic fatty carboxylic acid.	Good adhesion Good chemical resistance
Amine adduct	Adduct reaction of polyamine with epoxy resin.	Good reactivity Low toxicity
Phenalkamine	Mannich reaction of polyamine with phenol and aldehyde.	Fast curability Good chemical resistance

Polyamides-type Curing Agents

A polyamide-type curing agent is prepared by reacting dimeric acid with an excessive polyamine, as illustrated in Fig. (**18**) [42]. Polyamide has higher viscosity, lower toxicity, and longer pot life than polyamines. Polyamides show good adhesion, flexibility, and toughness. Typical unmodified polyamide-type epoxy curing agents are summarized in Table **16**.

Table 16. Typical polyamide-type curing agents.

AHEW, g/eq	Viscosity, mPa.s/25°C	Mix Ratio, phr*	Gel Time, min@25°C	Commercial Products**
100	2000-4000	50	170	Ancamide 375A, Versamid 150

(Table 16) cont.....

100-120	8000-14000	55	200	Ancamide 350A, Versamid 140, Epikure 3140.
120-140	8000-40000	65	120	Ancamide 260A, Versamide 125, Epikure 3125.
165-185	50000-400000	100-110	260	Ancamide 220, Versamide 115, Epikure 3115.

*phr: parts per hundred resin.

**Versamid products are supplied by Huntsman Corporation, Ancamide products are supplied by Evonik Corporation, Epikure products are supplied by Westlake Incorporation.

Fig. (18). Synthesis of polyamide from dimeric acid and diamine.

Amidoamine-type Curing Agents

Amidoamine-type curing agent is prepared by the reaction of a monobasic fatty carboxylic acid with an excessive polyamine. Amidoamine has a viscosity between polyamide and polyamine. Amidoamine-type curing agents show good handling properties, good adhesion, and good chemical resistance.

Amine Adduct-type Curing Agents

Amine adduct-type curing agent is prepared by an adduct reaction of epoxy resin with excessive polyamine at a well-controlled temperature of around 75°C, as illustrated in Fig. (**19**) [43]. Amine adducts offer good reactivity, low volatility, and a high mixing ratio for easy handling. The reason for good reactivity is because the amine adduct is already partially reacted with epoxy resin, and thus, less reaction is needed for gelation.

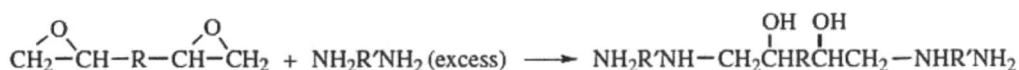

$$CH_2-CH-R-CH-CH_2 + NH_2R'NH_2 \text{ (excess)} \longrightarrow NH_2R'NH-CH_2\overset{OH}{\underset{|}{C}}HR\overset{OH}{\underset{|}{C}}HCH_2-NHR'NH_2$$

Fig. (19). Synthesis of amine adduct from epoxy resin and diamine.

Phenakamine-type Curing Agents

Phenalkamine is a Mannich reaction product of polyamine with phenol and aldehyde, as illustrated in Fig. (**20**) [44]. Phenalkamine offers fast curability and good chemical resistance. An internal phenolic accelerator contributes to its faster curability.

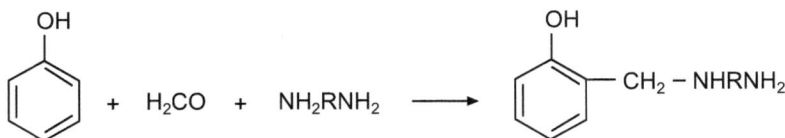

Fig. (20). Synthesis of phenalkamine *via* Mannich reaction.

Mercaptans

Mercaptans can cure epoxide very fast in the presence of a basic catalyst at room temperature. Mercaptans are often used as the main curing agents for fast room-temperature epoxy systems. The cure mechanism of epoxy resin and mercaptan in the existence of a tertiary amine catalyst is shown in Fig. (**21**) [45]. As a result, the epoxy group reacts with mercaptan almost equivalently to form alcohol and sulfide. The mercaptan or thiol group reacts with the epoxy group very slowly by itself in the absence of a catalyst at room temperature. Details will be described in Chapter 6.

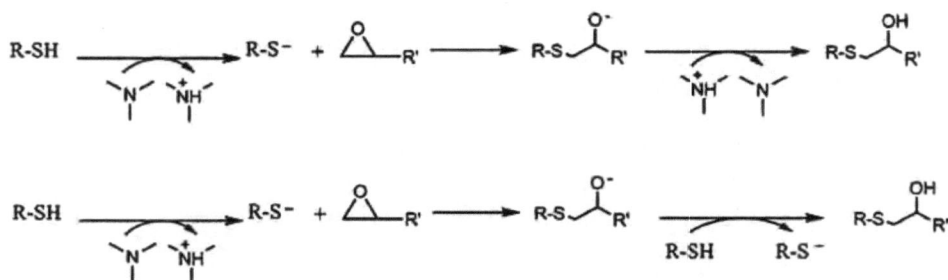

Fig. (21). Polyaddition mechanism of epoxy and mercaptan.

Thiol equivalent is thiol equivalent weight, representing weight in grams of thiol or mercaptan curing agent per one thiol group. Thiol equivalent is a very important parameter in epoxy resin technology. The calculation of the thiol equivalent is shown in Eq. **1.5**.

Thiol equivalent = Molar mass of mercaptan curing agent / Number of the thiol group

Equation 1.5 Thiol equivalent calculation.

Phenols

Phenols can react with epoxy resin in the existence of a small amount of catalyst almost equivalently at elevated temperatures, as illustrated in Fig. (23) [46]. The chemical structure and key features of typical phenol-type curing agents are shown in Fig. (24). Catalysts are TPP (triphenylphosphine), tertiary amine, or imidazole compounds. Phenol-cured epoxies are mainly used for electrical castings, encapsulants, or molding compounds.

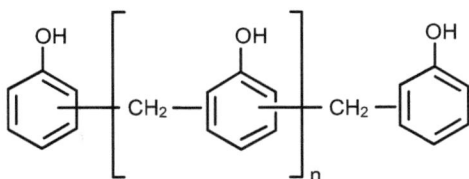

Phenol novolac resin

Softening point: 60 -120°C
CAS No.: 9003-35-4
OH equivalent: 103-106
Standard phenol type curing agent

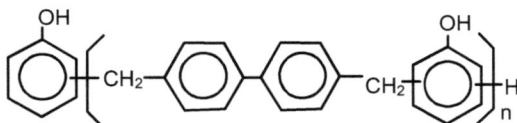

Biphenol novolac resin

Softening point: 60 -120°C
CAS No.: 26834-02-6
OH equivalent: 200-210
Flame retardancy

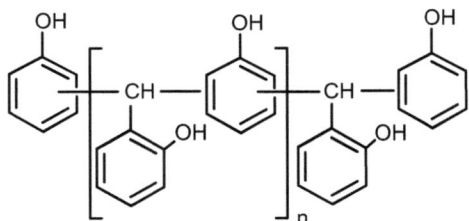

Triphenylmethane novolac resin

Softening point: 80 -120°C
CAS No.: 106466-55-1
OH equivalent: 97-103
High thermal resistance

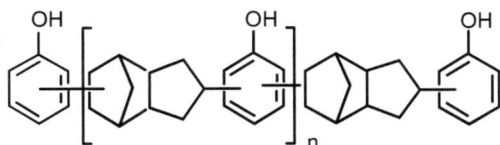

Dicyclopentadiene phenyl novolac resin

Softening point: 95 -112°C
CAS No.: 30420-31-6
COH equivalent: 170-180
Humidity resistance

Fig. (22). Typical phenol-type curing agents.

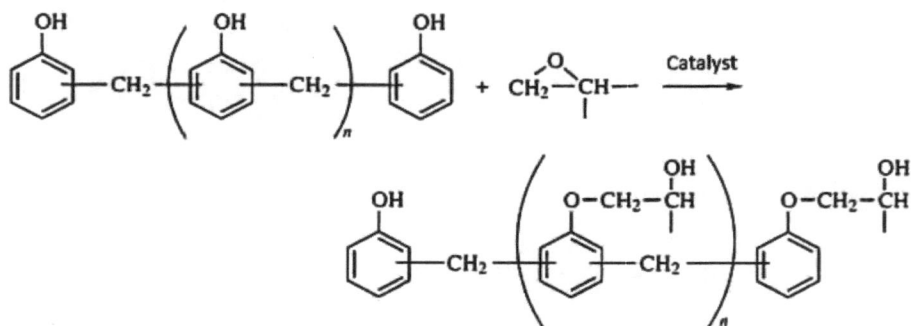

Fig. (23). Polyaddition mechanism of epoxy and phenols.

OH equivalent is OH equivalent weight, representing weight in grams of phenol-type curing agent per one -OH (phenol) group. The calculation of the OH equivalent is shown in Eq. **1.6**.

OH equivalent = Molecular weight of phenol type curing agent/Number of -OH group

Equation 1.6 OH equivalent calculation.

Anhydrides

Anhydrides have been used as curing agents for epoxy resin since the very beginning of epoxy resin technology development. The chemical structure and key features of typical commercial anhydride-type curing agents are shown in Fig. (**24**). The cure mechanism of anhydride for epoxy resin is very complicated due to several possible competing reactions. The reaction of anhydride alone with epoxy resin, as illustrated in Fig. (**25**) [47], is very slow below 200°C. Usually, catalysts such as tertiary amine, substituted imidazole, and TPP are combined together. Fig. (**26**) [48] shows the cure route of anhydride and epoxy resin catalyzed by a tertiary amine. Anhydride-formulated epoxy resin exhibits low viscosity, long pot life, low curing exotherm, and good physical and electrical properties, suitable for use in large mass curing applications such as casting, potting, and reinforced materials. Their use in adhesive applications is limited due to the slow reactivity and high cure temperature required.

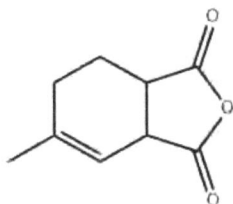

Viscosity: 30 -60 mPa.s/25°C
CAS No.: 42498-58-8
anhydride equivalent: 170
general purpose

Methyltetrahydrophthalic anhydride (MTHPA)

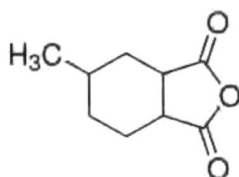

Viscosity: 50 -80 mPa.s/25°C
CAS No.: 48122-14-1
anhydride equivalent: 164
transparency

Methyl hexahydrophthalic anhydride (MHHPA)

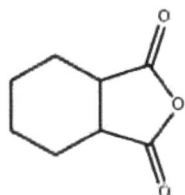

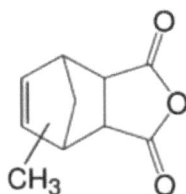

Melting point: 34°C
CAS No.: 85-42-7
anhydride equivalent: 154
transparency

Hexahydrophthalic anhydride (HHPA)

Viscosity: 1000 - 15000 mPa.s/25 °C
CAS No.: 53584-57-9
anhydride equivalent: 178
thermal resistance

Methyl himic anhydride (MHA)

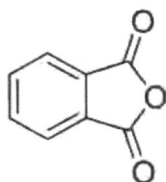

Melting point: 128°C
CAS No.: 85-44-9
anhydride equivalent: 178
inexpensive

Phthalic anhydride (PA)

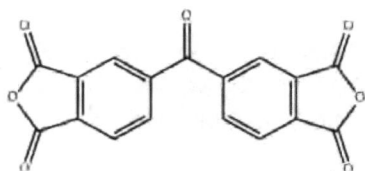

Melting point: 227°C
CAS No.: 2421-28-5
anhydride equivalent: 161
thermal resistance

Benzophenone tetracarboxylic dianhydride (BTDA)

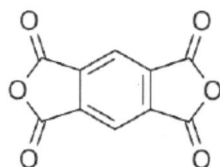

Melting point: 286°C
CAS No.: 89-32-7
anhydride equivalent: 109
thermal resistance

Pyromellitic dianhydride (PMDA)

Fig. (24). Typical anhydride type curing agents.

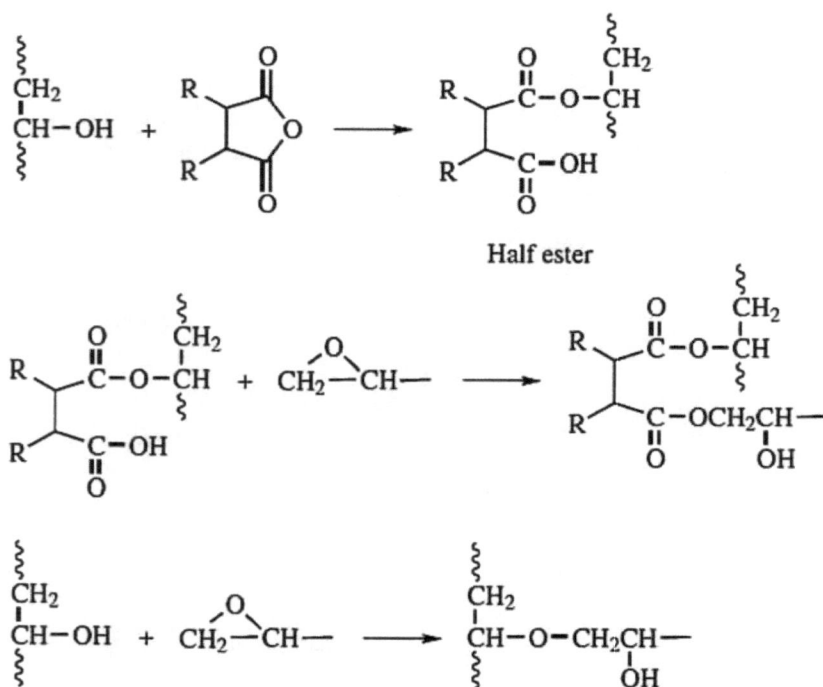

Half ester

Fig. (25). Reaction mechanism of epoxy resin and anhydride without catalyst.

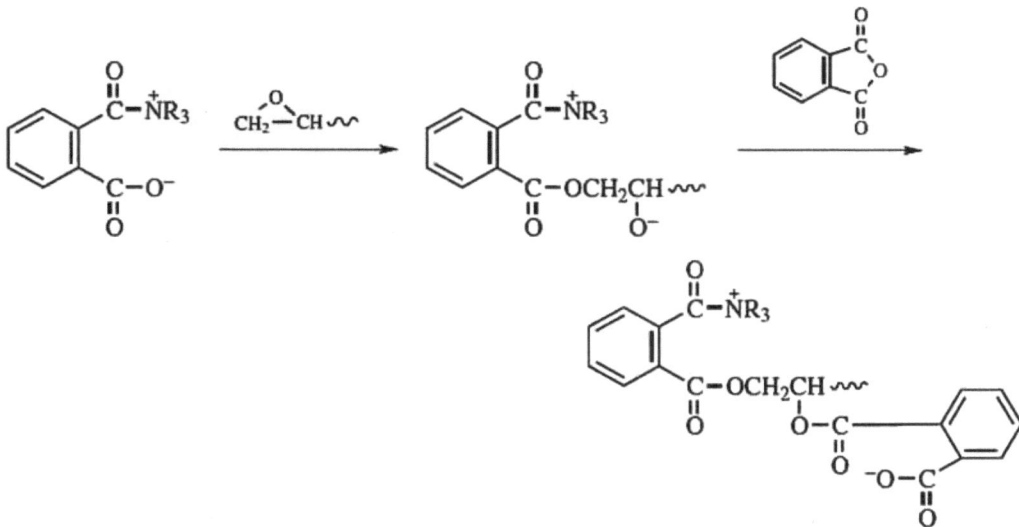

Fig. (26). Reaction mechanism of epoxy resin and anhydride catalyzed by tertiary amine.

Anhydride equivalent is anhydride equivalent weight, representing weight in grams of anhydride-type curing agent per one anhydride group. The calculation of the anhydride equivalent is shown in Eq. **1.7**.

Anhydride equivalent = Molar mass of anhydride type curing agent

/ Number of anhydride group

Equation 1.7. Anhydride equivalent calculation.

Tertiary Amines and Imidazoles

Tertiary amines and imidazole compounds are a type of Lewis base. They are often used as catalysts to cure epoxy resin *via* anionic polymerization. The reaction mechanism of tertiary amine-catalyzed anionic polymerization of epoxy resin is illustrated in Fig. (**27**). [49]. Tertiary amines and imidazole compounds are also used as accelerators for other curing agents, such as polyamines, mercaptans, anhydrides, and phenols. The chemical structure and key features of typical tertiary amine-type curing agents are shown in Fig. (**28**). Imidazole compounds will be described in Chapter 4.

$$-CH\!-\!CH_2 \;+\; R_3N \;\longrightarrow\; R_3N^+ \;\;-CH_2CH\!-\;\;\text{O}^-$$

Zwitterion

$$R_3N^+ \;\;-CH_2\overset{O^-}{CH}\!- \;+\; R'OH \;\longrightarrow\; R_3N^+\!-CH_2\overset{O^-}{CH}\!- \;+\; R'O^-$$

$$R'O^- \;+\; CH_2\!-\!CH\!- \;\longrightarrow\; R'O\!-\!CH_2\overset{O^-}{CH}$$

$$R'O\!-\!CH_2\overset{O^-}{CH} \;+\; n\,CH_2\!-\!CH \;\longrightarrow\; R'O\!\left(\!CH_2CHO\!\right)_{\!n}\!-\!CH_2\overset{O^-}{CH}$$

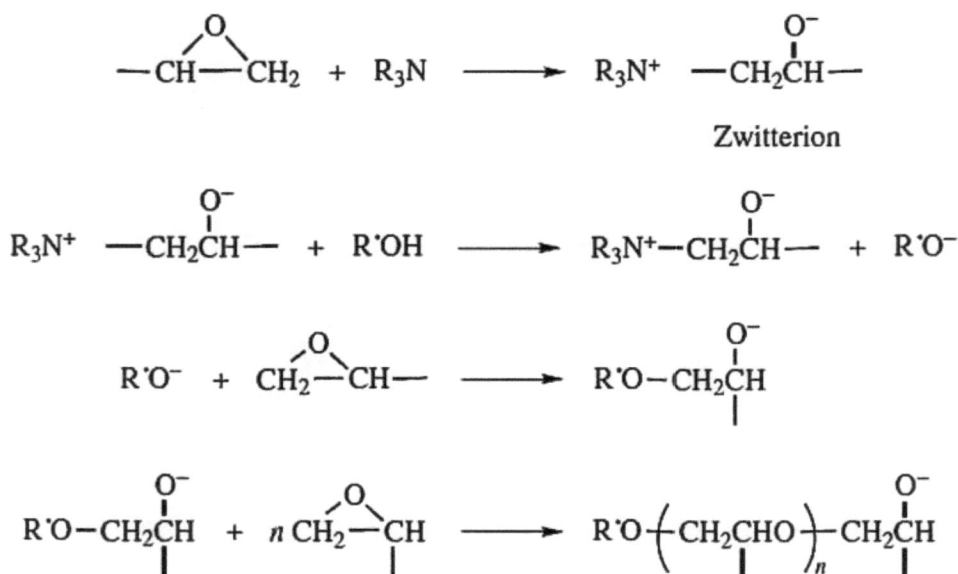

Fig. (27). Anionic polymerization of epoxy resin catalyzed by tertiary amine.

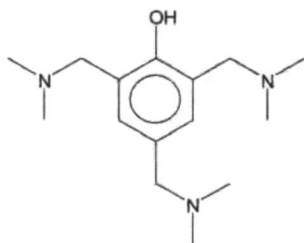

Viscosity: 120 -250 mPa.s/25°C
CAS No.: 90-72-2
Molar mass: 265

2,4,6-Tris(dimethylaminomethyl)phenol

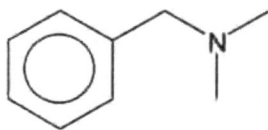

Viscosity: 10 mPa.s/25°C
CAS No.: 103-83-3
Molar mass: 135

BDMA (benzyldimethylamine)

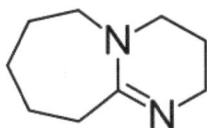

1,8-Diazabicyclo[5.4.0]undec-7-ene (DBU)

Light yellow liquid
CAS No.: 6674-22-2
Molar mass: 152

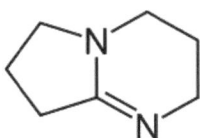

Light yellow liquid
CAS No.: 3001-72-7
Molar mass: 124

1,5-Diazabicyclo[4.3.0]non-5-ene (DBN)

Fig. (28). Chemical structures of typical tertiary amine curing agents.

Cationic Initiators

Boron trifluoride monoethylamine complex has been used as a latent curing agent for epoxy resin for a long time. Its pot life at room temperature is a few months at room temperature. As shown in Fig. (**29**) [50], the boron trifluoride monoethylamine complex reacts with the epoxy group at elevated temperatures to form cationic species that will initiate the cationic polymerization of epoxy resin. The complex is very sensitive to moisture. Precautions need to be taken to avoid basic additives.

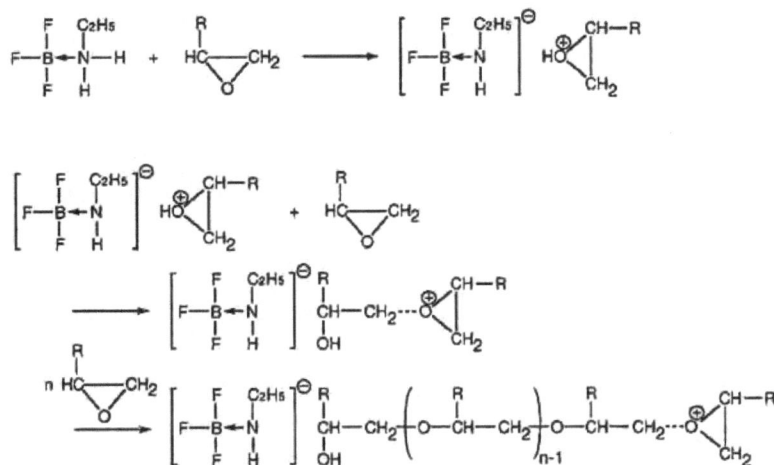

Fig. (29). Cationic polymerization of epoxy resin initiated by boron trifluoride monoethylamine complex.

Certain onium salts and iodonium salts will form strong acids *via* light radiation or by heating, initiating the curing of epoxy resin *via* the cationic polymerization mechanism. The typical structure of onium salts and iodonium salts for UV cationic epoxy systems is shown in Fig. **(30)**. UV cure cationic epoxy adhesives will be described in detail in Chapter 2.

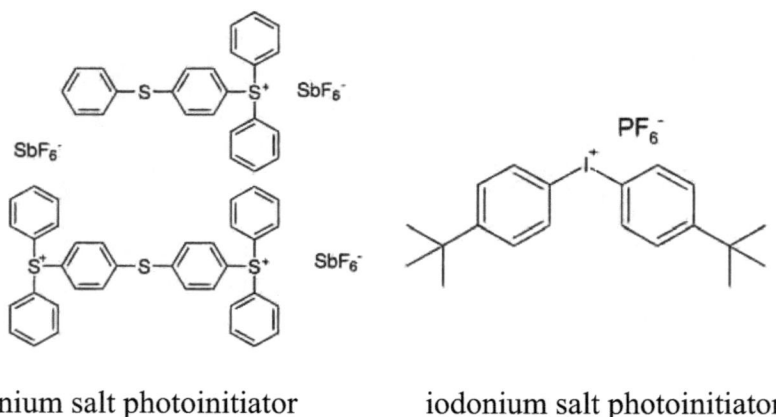

Onium salt photoinitiator   iodonium salt photoinitiator

Fig. (30). Chemical structures of onium salt and iodonium salt photoinitiators.

EPOXY ADHESIVES

Formulating Epoxy Adhesives

Epoxy adhesives are typically formulated from epoxy resin, curing agent, and catalyst with various modifiers and additives [51-54]. Key functions, ingredients, and their main role are summarized in Table **17**.

Table 17. Epoxy adhesive composition.

Function	Component	Main Role
Primary	Epoxy resin	Adhesive base
	Epoxy diluent	Viscosity adjustment
	Curing agent/catalyst	Curability, stability
	Accelerator	Cure speed enhancement

(Table 17) cont.....

Modifier	Filler	Property enhancement, cost down
	Toughener	Toughness enhancement
	Plasticizer	Flexibility enhancement
Additive	Colorant	Coloring
	Coupling agent	Adhesion promotion
	Thixotropic agent	Rheology control
	Others	

Epoxy adhesives can be formulated in either one-component or two-component packages by selecting a suitable curing agent/catalyst. Two-component epoxy adhesive is prepared by packaging epoxy composition, commonly called resin part or part A, and curing agent composition, commonly called hardener part or part B, separately before use. The two-component epoxy adhesives will start to cure after mixing the two parts together. One-component epoxy adhesive is prepared by mixing all ingredients, including epoxy resin and curing agent, in advance by selecting the use of suitable latent curing agents. One-component epoxy adhesives usually need thermal cure at elevated temperature conditions and require chilled or even frozen storage conditions at lower temperatures to ensure a long shelf life. UV cationic epoxy adhesives are normally prepared in a one-component package with the use of cationic photoinitiators that can generate cationic species *via* UV radiation, initiating quick cure of epoxy resin *via* cationic polymerization mechanism.

Below is a general process for designing a successful epoxy adhesive formulation:

1. Selection of epoxy resin, epoxy diluent.

2. Selection of curing agent, catalyst, and accelerator.

3. Selection of proper modifiers.

4. Selection of proper additives.

5. Calculation of epoxy resin and curing agent ratio.

Theoretically, a cross-linked cured epoxy structure can be obtained when equimolar quantities of epoxy resin and curing agent are formulated for polyaddition-type curing agents. The actual mixing ratio, however, will depend very much on the curing agent type. In polyamine-cured epoxy systems, the curing agent is formulated in a near stoichiometric ratio with epoxy resin. This is because tertiary amine formed during the curing process has a catalytic effect on the following curing process. Experiencedly, 0.9 to 1.0 of amine active hydrogen to 1 epoxy group is used. In anhydride-cured epoxy system, the curing agent is formulated in a less stoichiometric ratio with epoxy resin. Typically, 0.5 to 0.85 of anhydride to 1 epoxy group is used because of the occurrence of significant epoxy homopolymerization. For mercaptan and phenol-cured systems, the mix ratio is between these two cases. The additional amount of pure catalyst or cationic initiator becomes much smaller, typically 1 to 20 parts per 100 epoxy resin. It is determined mainly by experience and testing results. Typical thermal, physical, and mechanical properties, chemical resistance, and thermal degradation of standard liquid bisphenol A epoxy resin cured by polyamines, polyamide, anhydride, and Lewis acid type curing agents are compared in Table **18** [55].

Table 18. Typical properties of standard liquid DGEBA (EEW=185) cured with different curing agents.

-	TETA	MXDA	Polyamide[1]	MHHPA[2]	BF$_3$-MEA
Phr	13	26	43	87.5	3
Cure schedule, hours (°C)	16 (25) + 3 (100)	16 (55) + 2 (125)	16 (25) + 3 (100)	4 (100) + 4 (165)	4 (100) + 16 (150)
HDT, °C	111	160	101	156	168
Strength, MPa					
Compression	112	116	85.6	126	114
Flexural	96	93	67	97	100
Tensile	79	70.4	57.3	69	39.4

(Table 18) cont.....

Modulus, GPa					
Compression	3.05	2.6	2.6	3.04	2.3
Flexural	3.05	2.7	2.14	3.05	3.1
Elongation, %	4.4	4.4	3.9	2.5	1.6
Electrical resistance, $10^{-17}\Omega\cdot m$					
volume	6.1	12.2	12.2	6.1	8.6
surface	7.8	>7.9	5.5	>7.3	>7.9
Weight gain, % after 28 days@					
50% NaOH	0.04	-0.05	0.07	-0.12	-0.02
30% H_2SO_4	1.8	1.6	1.9	0.83	1.1
Acetone	2.1	4.6	7.3	15.0	1.2
Toluene	0.07	0.13	3.7	0.09	0.17
Water	0.86	1.1	1.3	0.82	1.2
Weight loss, % after 300h @ 210°C	6.8	5.5	5.0	1.5	4.9

1. Versamide 140.

2. Catalyzed with 1.5 phr of BDMA.

Fillers are often incorporated in epoxy adhesive formulations to reduce the cost and enhance or obtain specific desired properties such as mechanical and electrical properties. Fillers can be inorganic or organic, but most used are inorganic type. The shape of fillers can be spheroidal, granular, fibrous, or lamellar. Typical fillers, their chemical composition, supplied particle shape, density, and main function are summarized in Table **19**.

Table 19. Typical fillers.

Filler	Chemical Composition	Particle Shape	Density	Main Function
Calcium carbonate	$CaCO_3$	Crystalline	2.71	Extender
Silica	SiO_2	Spheroidal	2.65	Extender
Alumina	Al_2O_3	Spheroidal	3.95	Abrasion resistance, thermal conductivity
Barium sulfate	$BaSO_4$	Granular	4.50	Extender
Silica carbide	SiC	Granular	3.21	Abrasion resistance
Kaolin clay	$Al_2Si_2O_5(OH)_4$	Granular	2.65	Extender
Talc	$Mg_3Si_4O_{10}(OH)_2$	Platy	2.7	Extender
Fumed silica	SiO_2	Granular	2.1	Thixotropic agent
Titanium dioxide	TiO_2	Granular	4.23	Pigment
Glass fiber	$(SiO_2)_n$	Fibrous	2.6	Reinforcement
Aluminum trihydroxide	$Al(OH)_3$	Granular	2.42	Flame retardant
Aluminum	Al	Granular	2.7	Inductivity
Silver	Ag	Granular	10.49	Electrical conductivity

Calcium carbonate is the most commonly used filler in epoxy technology. It is widely available and inexpensive. Calcium carbonate can also be used to improve the mechanical properties of epoxy adhesives. Calcium carbonate includes limestone, marble, calcite, and chalk. They are supplied by various manufacturers in different particle sizes.

Silica is also a common filler used in epoxy technology, especially in semiconductor packaging and electronics assembly applications, because of its very low-level of CTE (coefficient of thermal expansion). Fumed silica is commonly used to provide thixotropic properties for epoxy adhesives.

Alumina and silicon carbide are used to enhance abrasion resistance and thermal conductivity in epoxy technology.

Silver is the most commonly used filler in electrically conductive epoxy adhesives.

Toughener is incorporated with epoxy adhesives to improve its resistance to failure tendency under mechanical stress. Various core-shell rubber tougheners have been commercialized recently. Core-shell rubber bead is typically very fine powder. Its core part is a synthetic rubber, while its shell part is a crosslinked polymer. Core-shell rubber tougheners can improve not only the toughness property significantly but also the adhesion strength and thermal shock resistance of epoxy adhesives. Unlike conventional flexible rubber tougheners, there is almost no obvious drop in glass transition temperature due to core-shell rubber's crosslinked shell structure. The appearance, microscopy picture, and structure of a Kana Ace CSR premix are shown in Fig. (**31**). Core-shell rubber tougheners containing epoxy adhesives have been increasingly used in high-performance structural bonding applications [56-62].

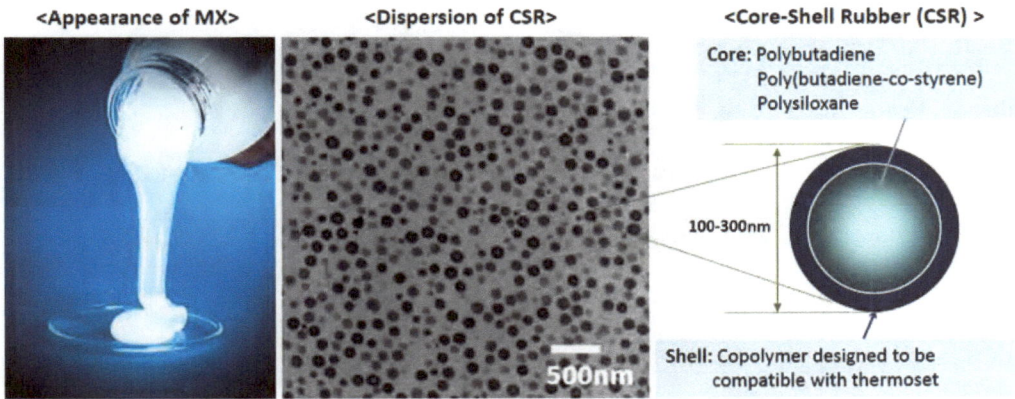

Fig. (31). Appearance, microscopy picture, and structure of Kane Ace CSR premix.

Silane coupling agent is the most commonly used adhesion promoter in epoxy adhesive technology. Fig. (**32**) shows the chemical structure and key features of typical epoxy silane coupling agents. 3-(Glycidyoxypropyl)trimethoxysilane is widely used in epoxy adhesive technology to improve adhesion on various inorganic substrates. Trimethoxy[2-(7-oxabicyclo[4.1.0]hept-3-yl)ethyl]silane can be used for cationic epoxy formulations since it has a cycloaliphatic epoxy group.

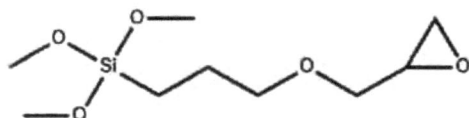

Clear liquid
CAS No.: 2530-83-8
Molar mass: 236

3-(Glycidoxypropyl)trimethoxysilane

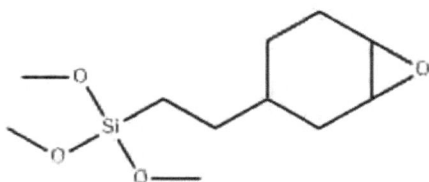

Clear liquid
CAS No.: 3388-04-3
Molar mas: 246

Trimethoxy[2-(7-oxabicyclo[4.1.0]hept-3-yl)ethyl]silane

Fig. (32). Typical epoxy silane coupling agents.

Properties, Testing, and Characterization

Typical properties, standard test methods, and test equipment for epoxy adhesives are summarized in Table **20** [63-66].

Table 20. Properties, testing, and characterization of epoxy adhesives.

Stage	Property	Standard Test Method	Test Equipment
Uncured state	Appearance	ASTM E 284-17	Visual inspection
	Viscosity	ISO 3219, ASTM D2196	Viscometer
Curing stage	Pot-life	ISO 10364, ASTM 2471	Viscometer
	Gel time	ISO 2535, ASTM D2471	DSC
	Cure behavior	ISO 11357, ASTM E 968	Density meter
	Cure shrinkage	ISO 1675, ASTM D 792	
Cured adhesive	Hardness	ISO 868, ASTM 2240	Hardness gauge

(Table 20) cont.....

	Modulus	ISO 527-3, ASTM D 882	Tension tester
	CTE	ISO 11359-2, ASTM E 1545	TMA
	Tg	ISO 11359-2, ASTM E 1545	DSC, TMA, DMA
Adhesion performance	Shear strength	ISO 4587, ASTM 1002-94	Tension tester
	Tensile strength	ISO 6922, ASTM D2095-92	Tension tester
	Peel strength	ISO 11339, ASTM D1876-93	Tension tester
	Impact strength	ISO 9653, ASTM 950-94	Impact tester

Typical Epoxy Adhesives

Room Temperature Cure Epoxy Adhesives

Room temperature cure epoxy adhesives are typically supplied in two-component packages. As shown in Table **21**, room-temperature cure epoxy adhesives can be further classified as fast-cure, general-purpose, and high-performance epoxy adhesives in actual uses.

Table 21. Typical room temperature cure epoxy adhesives.

Property	Fast Cure	General-Purpose	High-Performance
Mix ratio, resin/hardener	1:1, 2:1		
Color	Clear, white, grey, black		White, grey, black
Mix viscosity, mPa.s/25°C	5000 - 300000		
Pot-life, min @25°C	<10	20 - 300	
Fixture time, min @25°C	< 20	60 - 300	
Full cure time, days	1	3 - 7	

(Table 21) cont.....

Lap shear strength, MPa on Al	15 - 25		>25
Max operating temperature, °C	80		120

There are various types of containers used for two-component epoxy adhesives depending on the application and usage amounts. For easy use, two-component epoxy adhesives are often packed in a well-designed 2K cartridge and can be simply mixed and dispensed with using a static mixer, as shown in Fig. (33).

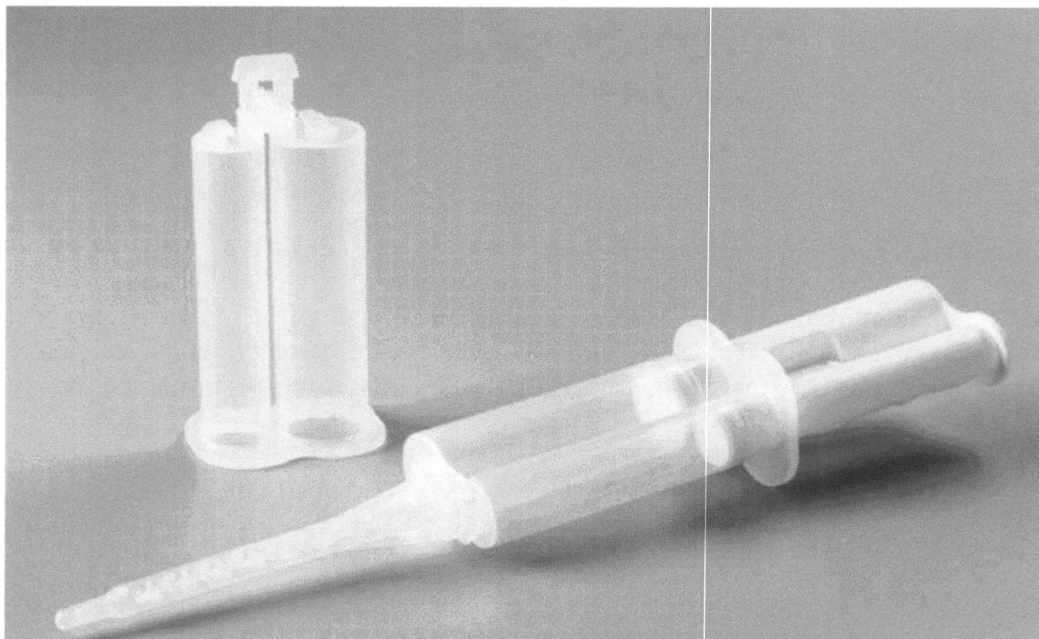

Fig. (33). 2K cartridge and static mixer.

Fast Room Temperature Cure Epoxy Adhesives

Mercaptan is typically selected as the main curing agent for fast room-temperature cure epoxy systems. Mercaptans can react with epoxy resin very fast at room temperature conditions in the presence of a small amount of basic chemicals *via* the polyaddition reaction mechanism of the epoxy group and thiol group. Gel time can be very short, less than 15 minutes or even 10 minutes. Precautions need to be taken into consideration in actual uses because their pot life is very short. Tg of mercaptan-based epoxy adhesives is not high, usually less than 50°C due to its aliphatic structure base, and thus their thermal resistance is relatively limited. Gel-time, tack-free time, and hardness generation of two simple mercaptan epoxy resin compositions are shown in Table **22**. As can be seen, both the tertiary amine and polyamide-catalyzed mercaptan-epoxy resin compositions showed fast curability not only at room temperature but also at a relatively low temperature of 5°C.

Table 22. Fast room-temperature cure epoxy compositions.

Resin component		
Epoxy resin[1]	100	100
Curing agent component	80	75
Mercaptan curing agent[2]	-	25
Polyamide	1	-
2,4,6-tris(dimethylaminomethyl)phenol		
Gel-time, min		
@5 °C	40	7.5
@23 °C	14	4
Tack Free Time, min		
@5 °C	24	5
@23 °C	24	4
Hardness, Shore D		
15min @ @23 °C	2	20
30 min	3	40
60 min	7	50
3 hours	9	57

(Table 22) cont.....

24 hours	10	63
7 days	40	70
Hardness, Shore D		
15min @ @5 °C	0	0
30 min	0	7
60 min	13	14
3 hours	33	23
24 hours	35	45
7 days	45	60

1. jER 828 (EEW=190) manufactured by Mitsubishi Chemical Corporation.

2. QE-340 (Thiol content: 12%) manufactured by Toray Fine Chemicals Co., Ltd.

One commercial fast-cure epoxy adhesive supplied by Henkel Corporation and its typical properties are illustrated in Table **23** [67].

Table 23. Typical properties of fast-cure epoxy adhesive product.

Product	LOCTITE EA E-00CL
Chemical type	Epoxy resin/mercaptan
Viscosity, mPa.s/25°C	
Resin	7000 - 13000
Hardener	2200 - 4500
Specific gravity, @25°C	
Resin	1.1
Hardener	1.1

(Table 23) cont.....

Appearance	Clear liquid
Mix ratio, by volume	1:1
Work life, minutes @25°C	3.5
Fixture time, minutes @25°C	4.0
Room temperature cure time, day	1
Glass transition temperature, °C	20
Tensile strength, MPa	26
Elongation at break, %	13
Shear strength, MPa	
Steel (grit blasted)	15
Stainless steel	14
Aluminum (anodized)	13
Polycarbonate	6.9
Nylon	1.2

General Purpose Epoxy Adhesives

Aliphatic polyamines and modified polyamines are the most commonly used curing agents in epoxy resin technology. Viscosity, cure behavior, physical property, and chemical property of aliphatic polyamine and modified polyamine are compared in Table **24**.

Table 24. Property comparison of aliphatic and modified polyamines.

-	Aliphatic Polyamine	Amine Adduct	Phenalkamine	Polyamide	Amidoamine
Viscosity	low	moderate	moderate	high	Moderate

(Table 24) cont.....

Addition amount	small	moderate	moderate	high	moderate
Pot-life	short	short	short	long	long
Cure speed	fast	fast	Very fast	slow	slow
Tensile strength	high	high	high	low	low
Peel strength	poor	poor	poor	high	High
Thermal resistance	good	good	good	poor	poor
Moisture resistance	good	good	good	poor	Poor
Hazardous vapor	high	moderate	moderate	low	moderate

Fixture time and work life can be adjusted by combining with a suitable curing agent. Table **25** shows one commercial room-temperature epoxy adhesive supplied by Henkel Corporation and its properties [68].

Table 25. Typical properties of general-purpose epoxy adhesive product.

Product	LOCTITE EA E-30CL
Chemical type	Epoxy resin/amine
Viscosity, mPa.s/25°C	
Resin	10,500
Hardener	2,250
Specific gravity, @25°C	
Resin	1.1
Hardener	1.0

(Table 25) cont.....

Appearance	Clear liquid
Mix ratio, by volume	1:1
Work life, minutes @25°C	30
Fixture time, minutes @25°C	160
Room temperature cure time, days	5
Glass transition temperature, °C	61
Volume shrinkage, %	4.3
Tensile strength, MPa	55
Elongation at break, %	8
Shear strength, MPa	
Steel (grit blasted)	21
Stainless steel	21
Aluminum (anodized)	14
Polycarbonate	13
Nylon	2.4

High-performance Epoxy Adhesives

Room temperature cure two-component high-performance epoxy adhesive is usually prepared by incorporating it with toughening technology. The high-performance epoxy adhesive possesses higher adhesion performance than general-purpose epoxy adhesive and can be widely used for various structural bonding applications requiring higher reliability and durability. Table **26** shows the typical properties of a high-performance epoxy adhesive product supplied by Henkel Corporation [69].

Table 26. Typical properties of high-performance epoxy adhesive product.

Product	LOCTITE EA E-60HP
Chemical type	Epoxy resin/amine
Viscosity, mPa.s/25°C	
Resin	67,500
Hardener	5,000
Specific gravity, @25°C	
Resin	1.0
Hardener	1.0
Appearance	Off-white liquid
Mix ratio, by volume	1:1
Work life, minutes @25°C	60
Fixture time, minutes @25°C	120
Room temperature cure time, days	5
Glass transition temperature, °C	70
Tensile strength, MPa	35
Elongation at break, %	9
Shear strength, MPa	
Steel (grit blasted)	30
Stainless steel	27
Aluminum (anodized)	30
Polycarbonate	13
Nylon	1.9

Thermal Cure Epoxy Adhesives

Two-component Thermal Cure Epoxy Adhesives

Two-component thermal cure epoxy adhesives are usually formulated by using cycloaliphatic amine or aromatic amine as a curing agent. Two-component thermal cure epoxy adhesives offer very good thermal and mechanical properties after full cure at elevated temperatures and are suitable for use in automobile production and aerospace assembly applications requiring high-temperature resistance. Table **27** shows the very good high-temperature resistant performance of a two-component epoxy adhesive incorporated with multifunctional epoxy resin and cycloaliphatic amine [70]. As can be seen, the epoxy adhesive can maintain very good adhesion performance at a very high temperature of 220°C.

Table 27. High-temperature resistant two-component epoxy composition.

Composition, %	
Resin part	98
Kana Ace 414[1]	2
Silane coupling agent	65
Hardener part	35
Ancamine 2264[2]	
Modified polyamine	
Mix ratio, by weight	70:30
Pot-life	60
Gel time, min at 23°C (10g scale)	120
Fixture time, min at 23°C	210
Lap shear strength, MPa/steel	
Cured at room temperature for 7 days	21.7
measured @23°C	12.0

(Table 27) cont.....

100	8.4
150	6.7
180	7.3
220	

1. N,N,N',N'-tetraglycidyl-4,4'-methylenebisbenzenamine based epoxy resin dispersed with 25% core shell beads, supplied by Kaneka Corporation.

2. Modified cycloaliphatic amine curing agent, supplied by Evonik Industries AG.

One-component Thermal Cure Epoxy Adhesives

One-component epoxy adhesives are prepared by selecting latent curing agents. All ingredients, including epoxy resin and curing agent, are mixed in advance. No additional pre-mixing process is required in actual use. There is no concern for insufficient mixing problems as often encountered in two-component uses. One-component adhesives can be handled easily and are suitable for automatic dispensing systems because of their long pot life. In recent years, one-component thermal cure epoxy adhesives have been increasingly used. One-component thermal cure epoxy adhesives normally need chilled storage conditions at lower temperatures. Many one-component thermal cure epoxy adhesives have been developed and supplied by epoxy adhesive suppliers for use as suitable latent curing agents. Typical commercial latent curing agents are summarized in Table **28**.

Table 28. Typical commercial latent curing agents.

Latent Curing Agent	Latency Mechanism	Curing Agent State	Typical Curing Temperature
DICY	Chemical blocking/physical separation	Solid	$\geqq$150°C
Dihydrazines			$\geqq$120°C
Modified imidazoles	Physical separation	Fine powder	$\geqq$80°C
Modified polyamine			$\geqq$80°C

(Table 28) cont.....

Onium salts	Chemical blocking	Solid	$\geqq 80°C$
Amine-BF$_3$ complex		Liquid	$\geqq 130°C$

DICY (dicyandiamide) is the most conventional latent curing agent for epoxy resin technology. Its chemical structure is shown in Fig. (**34**). DICY-formulated epoxy composition has very good stability, with more than 6 months at room temperature. The cured epoxy resin offers high adhesion and high thermal resistance. The cure temperature of DICY alone with epoxy resin is quite high. Accelerators such as modified urea compounds and imidazole compounds are often combined together to lower its cure temperature [71].

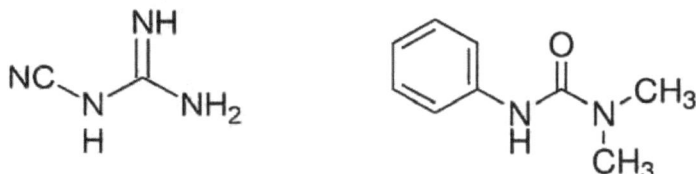

Fig. (34). Chemical structures of DICY (left) and substituted urea (right) accelerator.

Epoxy Film Adhesives

Epoxy adhesives are usually prepared and supplied in a liquid or paste state. They can also be prepared and supplied in the form of solid-state, typically as a film or tape. The epoxy adhesive film is normally a one-component type, processed to a solid/film stage but still not cured completely. When the applied adhesive film is heated, it is melted and cured completely. The epoxy adhesive film is typically prepared using the two methods mentioned below. Epoxy film adhesive with a release sheet is illustrated in Fig. (**35**).

A. All components, including epoxy resin and curing agent, are mixed in solvent to prepare a solvent epoxy solution that is then applied on a carrier sheet. The epoxy adhesive film is prepared by evaporating all the solvents.

B. All components, including epoxy resin and curing agent, are mixed to prepare a liquid epoxy composition that is then applied on a carrier sheet. The epoxy adhesive film is prepared by partially curing at certain curing conditions such as heating and light radiation.

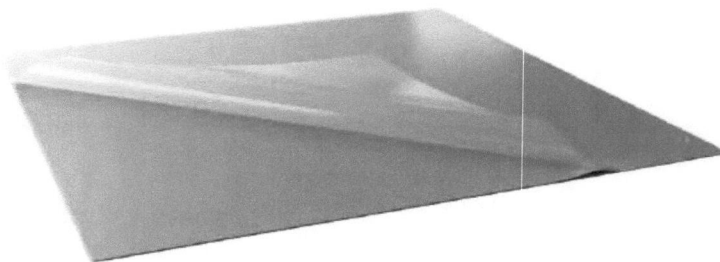

Fig. (35). Epoxy film adhesive with release sheet.

Table **29** shows the starting formulation of one general-purpose epoxy adhesive film. The adhesive film is prepared by mixing all the components into varnish and b-staging the varnish on a desired film backing material, followed by removing the solvent at 83°C for 1 hour. The typical cure condition of this epoxy adhesive film is 1 hour at 175°C with 120 psi pressure [72].

Table 29. Starting formulation of general-purpose adhesive film.

-	Parts by Weight
Solid epoxy resin (40% solids)	212
Epoxy novolac resin (85% solid)	37.5
Dicyandiamide	12
BDMA	2

UV Cure Epoxy Adhesives

UV (Ultraviolet light) cure epoxy technology includes UV cure cationic epoxy, UV and thermal dual cure hybrid epoxy, and UV cure anionic epoxy technology. UV cure cationic epoxy adhesives are primarily composed of epoxy resin and cationic photoinitiators with additives [73-75]. Cycloaliphatic type epoxy resins are typically selected for use in UV cationic epoxy technology as they can cure faster *via* cationic polymerization than normal glycidyl ether type epoxy resin. Cationic photoinitiators can absorb UV energy to form cationic species that can initiate homo-polymerization of epoxy resin. UV cationic epoxy adhesives have lower cure shrinkage, no surface cure issue resulting from oxygen inhibition, better adhesion, and shadow curability as compared to UV cure acrylate adhesives. UV cationic epoxy adhesives have been used successfully in optical parts bonding, camera module sensor packaging, and OLED (organic light-emitting diode) panel assembly applications [76-78]. UV and thermal dual cure hybrid epoxy adhesives combine UV cure acrylate composition with thermal cure epoxy composition. Dual cure hybrid epoxy adhesives are primarily composed of partially acrylated epoxy resin, acrylates, free radical photoinitiators, epoxy resin, and latent curing agents. Their acrylate components can be cured instantly *via* light radiation, and thus, the bonding materials can be fixed momently at room temperature. Their epoxy components are cured at the post-thermal cure stage to achieve satisfactory adhesion reliability performance. Dual cure hybrid epoxy technology has been very successfully used in electronics assembly and general bonding applications such as general structural bonding, camera module assembly, and LCD (liquid crystal display) panel bonding that require both fast production speed and high adhesion performance [79-80]. UV cure anionic epoxy technology is based on PBG (Photobase Generator) material, which can generate a strong base *via* UV radiation to initiate anionic polymerization of epoxy resin or accelerate the polyaddition reaction of epoxy resin with curing agents such as mercaptans [81-84].

Sustainable Epoxy Adhesive Development

Sustainable epoxy adhesive development primarily focuses on health and safety improvement, energy saving, and circular economy contribution [85-89]. Research and development on sustainable epoxy resins, curing agents, additives, and epoxy adhesives have become more and more important recently.

There are certain health and safety concerns for some raw materials used in epoxy adhesives, such as low-viscosity diluents and amine curing agents. Health and safety-friendly sustainable epoxy resins are developed by carefully selecting green raw materials and optimizing the formulation [90-92].

Sustainable biobased epoxy adhesives are formulated by selecting biobased raw materials. At the current stage, most biobased epoxies are prepared from the epoxidation of renewable precursors. Typical commercial bio-based epoxy resins are epoxidized linseed oil, liquid epoxidized natural rubber, terpene-maleic ester type epoxy, diglycidyl ether of isosorbide, and epoxidized cardanol. Typical biobased curing agents available commercially or under commercial investigation are vegetable oil-derived polyamides, cardanol-derived polyamines, polysaccharide or sugar-derived polyamines, and terpene-based anhydrides [93-111]. Fig. (**36**) illustrates the chemical structures of typical biobased epoxy resin, curing agent, and additive: aliphatic multifunctional epoxy resin based on sorbitol, bis(aminomethyl)furan based on polysaccharides and sugars, and cardanol, the main component of cashew nutshell liquid.

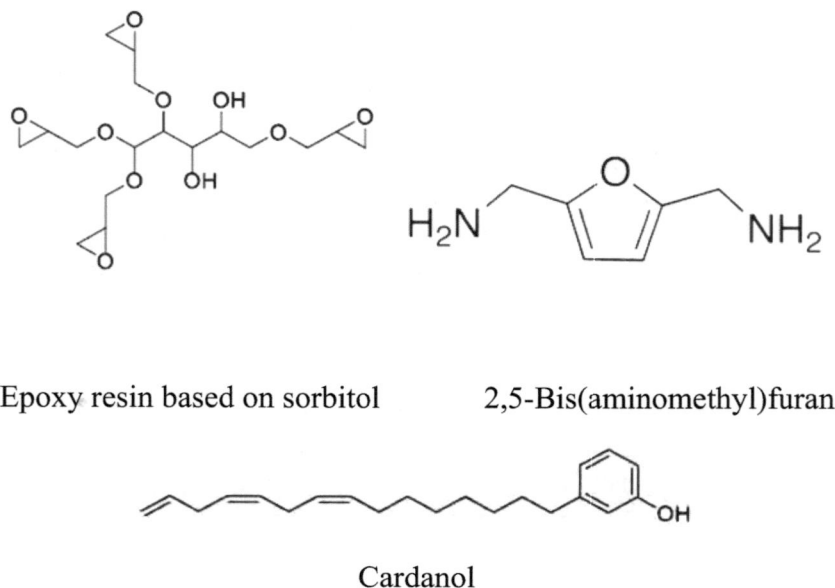

Epoxy resin based on sorbitol   2,5-Bis(aminomethyl)furan

Cardanol

Fig. (36). Typical biobased epoxy resin, curing agent, and additive.

Instant Bonding Epoxy Adhesive Introduction

A relatively long cure time, ranging from several days at room temperature to at least tens of minutes at elevated temperature, is usually required to cure epoxy adhesives. Recently, several new types of epoxy adhesives that can bond adherends instantly at the specified curing condition while still possessing satisfactory adhesion performance after full cure have been developed and successfully used in advanced applications such as general bonding, semiconductor packaging, and electronics assembly. Typical instant bonding epoxy adhesives are UV cure cationic epoxy adhesive, dual cure hybrid epoxy adhesive, snap thermal cure epoxy adhesive, induction cure epoxy adhesive, laser cure epoxy adhesive, weld bonding epoxy adhesive, and snap ambient cure epoxy adhesives. Their chemistry, instant bonding mechanism, cure equipment, and typical applications are compared in Table **30**. Fundamental theory and application knowledge of these typical instant epoxy adhesive technologies will be described in other chapters.

Table 30. Typical instant bonding epoxy adhesives.

Epoxy Adhesive Type	Chemistry	Instant Bonding Mechanism	Equipment	Typical Applications
UV cure cationic epoxy adhesive	Cationic epoxy	UV cure of epoxy	UV cure equipment	Electronics assembly; general purpose bonding
Dual cure hybrid epoxy adhesive	UV acrylate + thermal epoxy hybrid	UV cure of acrylate	UV cure equipment; conventional heating oven	Camera module assembly; LCD panel bonding
Snap thermal cure epoxy adhesive	Thermal cure epoxy	Snap thermal cure epoxy at elevated temperature	Hot press	Semiconductor packaging; electronics assembly
Induction cure epoxy adhesive	Thermal cure epoxy + susceptor		Induction cure equipment	Magnet bonding; hem flange bonding
Laser cure epoxy adhesive	Thermal cure epoxy + photothermal conversion material		Laser cure equipment	Electronics assembly
Weld bonding epoxy adhesive	Weld + Epoxy	Weld process	Weld	Aircraft assembly Automobile production
Snap ambient cure epoxy adhesives	Two-component epoxy	Snap room temperature cure epoxy	n/a	General purpose bonding
	Two-component cyanoacrylate epoxy hybrid	Anionic polymerization of cyanoacrylate	n/a	

(Table 30) cont.....

Two-component UV acrylate epoxy hybrid	UV cure of acrylate	UV cure equipment	

CONCLUSION

Epoxy adhesives are composed of epoxy resin, curing agent, and catalyst with modifiers and additives. Epoxy adhesives can be cured at room temperature condition, at elevated temperature condition, or *via* UV light radiation, mainly depending on the curing agent formulated. Lots of epoxy adhesives, either supplied in one-component or two-component packages, have been commercialized and widely used for bonding metals, , glass, ceramics, concrete, many plastics, wood, *etc.*, in various applications.

REFERENCES

[1]   May, C.A. *Epoxy Resins Chemistry and Technology,* 2nd ed; Marcel Dekker: New York, **1988**, pp. 1-8.

[2]   Pham, H.Q.; Marks, M.J. Epoxy resins. In: *Encyclopedia of Polymer Science and Technology*; Wiley: New York, **2004**, Vol. 13, pp. 155-244.
        http://dx.doi.org/10.1002/0471440264.pst119

[3]   Petrie, E.M. *Handbook of Adhesives and Sealants*; MicGraw-Hill: New York, **2006**, pp. 355-373.

[4]   Chen, C. Structural epoxy adhesives In: *Structural Adhesives: Properties, Characterization and Applications*; Mittal, K.L. and Panigraphi, S.K., Eds.; Wiley – Scrivener: Beverly, MA, **2023**, pp. 3-30.
        http://dx.doi.org/10.1002/9781394175604.ch1

[5]   Chen, C.; Li, B.; Kanari, M.; Lu, D. Epoxy adhesives. In: *Adhesives and Adhesive Joints in Industry Applications*; Rudawska, A., Ed.; IntechOpen: London, UK, **2019**, pp. 37-49.
        http://dx.doi.org/10.5772/intechopen.86387

[6]   Panda, H. *Epoxy Resins Technology Handbook*; Asia Pacific Business Press: New Delhi, **2016**.

[7]   Goulding, T.M. Epoxy resin adhesives. In: *Handbook of Adhesive Technology,* 2nd ed; Pizzi, A.; Mittal, K.L., Eds.; CRC Press: Boca Raton, FL, **2003**, pp. 809-824.

[8]   Bauer, M.; Schneider, J. Adhesives in the electronics industry. In: *Handbook of Adhesive Technology,* 2nd ed; Pizzi, A.; Mittal, K.L., Eds.; CRC Press: Boca Raton, **2003**, pp. 875-886.

[9]   Cordes, E.H. Adhesives in the automotive industry. In: *Handbook of Adhesive Technology,* 2nd ed; Pizzi, A.; Mittal, K.L., Eds.; CRC Press: Boca Raton, **2003**, pp. 983-999.

[10]   Rudawska, A. Epoxy adhesives. In: *Handbook of Adhesive Technology,* 3rd ed; Pizzi, A.; Mittal, K.L., Eds.; CRC Press: Boca Raton, **2018**, pp. 415-442.

[11]   Sancaktar, E.; Bai, L. Electrically conductive epoxy adhesives. *Polymers (Basel),* **2011**, *3*(1), 427-466.
http://dx.doi.org/10.3390/polym3010427

[12]   Severijns, C.; de Freitas, S.T.; Poulis, J.A. Susceptor-assisted induction curing behaviour of a two component epoxy paste adhesive for aerospace applications. *Int. J. Adhes. Adhes.,* **2017**, *75*, 155-164.
http://dx.doi.org/10.1016/j.ijadhadh.2017.03.005

[13]   Chen, C. Epoxy adhesive technology: latest development and new trends. In: *Progress in Adhesion and Adhesives Volume 8; Mittal*, K.L., Eds.; Wiley – Scrivener: Beverly, MA, **2024**, pp. 251-282.
http://dx.doi.org/ 10.1002/9781394238231.ch6

[14]   Zotti, A.; Zuppolini, S.; Zarrelli, M.; Borriello, A. Fracture toughening mechanisms in epoxy adhesives. In: *Adhesives – Applications and Properties*; Rudawska, A., Ed.; IntechOpen: London, **2016**, pp. 237-269.
http://dx.doi.org/10.5772/65250

[15]   Kwakernaak, A.; Hofstede, J.; Poulis, J.; Benedictus, R. Improvements in bonding metals for aerospace and other applications. In: *Welding and Joining of Aerospace Materials*; Chaturvedi, M., Ed.; Woodhead Publishing: Oxford, UK, **2012**; pp. 235-319.
http://dx.doi.org/10.1533/9780857095169.2.235

[16]   Ebnesajjad, S.; Landrock, A.H. *Adhesives Technology Handbook,* 3rd ed; Elsevier: London, **2015**, pp. 110-113.

[17]   Kucklick, T.R. *The Medical Device R&D Handbook*; CRC Press: Boca Raton, FL, **2013**, pp. 34-35.

[18]   Lewis, A.F. Epoxy resin adhesives. In: *Epoxy Resins – Chemistry and Technology,* 2nd ed; May, C.A., Ed.; Marcel Dekker: New York, **1988**, pp. 653-718.

[19]   Jin, F.L.; Li, X.; Park, S.J. Synthesis and application of epoxy resins: A review. *J. Ind. Eng. Chem.,* **2015**, *29*, 1-11.
http://dx.doi.org/10.1016/j.jiec.2015.03.026

[20]   Wright, C.D.; Muggee, J.G. Epoxy structural adhesives. In: *Structural Adhesives Chemistry and Technology*; Hartshorn, S.R., Ed.; Springer: Boston, **1988**, pp. 113-179.

[21]   Licari, J.J.; Swanson, D.W. *Adhesive Technology for Electronics Applications: Materials, Processing, Reliability,* 2nd ed; Elsevier: London, **2011**, pp. 77-87.

[22]   Jojibabu, P.; Zhang, Y.X.; Prusty, B.G. A review of research advances in epoxy-based nanocomposites as adhesive materials. *Int. J. Adhes. Adhes.,* **2020**, *96*, 102454.
http://dx.doi.org/10.1016/j.ijadhadh.2019.102454

[23]   Kakiuchi, H. *Epoxy Resin: A Review*; The Japan Society of Epoxy Resin Technology: Tokyo, **2003**.

[24]   Wahab, M.A. *Joining Composites with Adhesives*; DEStech: Lancaster, PA, **2016**.

[25]   Abbey, K.J. Advances in epoxy adhesives. In: *Advances in Structural Adhesive Bonding*, Dillard, D.A., Ed.; Woodhead: Cambridge, UK, **2010**, pp. 20-34.

http://dx.doi.org/10.1533/9781845698058.1.20

[26]   Liu, J.Q.; Bai, C.; Jia, D.D.; Liu, W.L.; He, F.Y.; Liu, Q.Z.; Yao, J.S.; Wang, X.Q.; Wu, Y.Z. Design and fabrication of a novel superhydrophobic surface based on a copolymer of styrene and bisphenol A diglycidyl ether monoacrylate. *RSC Advances,* **2014**, *4*(35), 18025-18032.
       http://dx.doi.org/10.1039/C4RA01505C

[27]   Muroi, S.; Ishimura, H. *Epoxy Resin Introduction*; Polymer Publishing Association: Tokyo, **1988**, pp. 2-3.

[28]   International Electrotechnical Commission. *Materials for Printed Boards and Other Interconnecting Structures – Part 2-21: Reinforced Base Materials, Clad and Unclad – Non-Halogenated Epoxide Woven E-Glass Reinforced Laminated Sheets of Defined Flammability (Vertical Burning Test)*, Copper-Clad; IEC Standard 61249-2-21, **2003**.

[29]   *Encyclopedia of chemical technology,* 5th ed; Wiley: New York, **2005**, Vol. 10, p. 381.

[30]   *Encyclopedia of chemical technology,* 5th ed; Wiley: New York, **2005**, Vol. 10, p. 378.

[31]   Takai, H. Other epoxy resins. In: *Epoxy resins*; Kakiuchi, H., Ed.; The Japan Epoxy Technology Society: Tokyo, **2003**, Vol. 1, pp. 76-84.

[32]   Kinloch, A.J. Toughening epoxy adhesives to meet today's challenges. *MRS Bull.,* **2003**, *28*(6), 445-448.
       http://dx.doi.org/10.1557/mrs2003.126

[33]   Stamper, D.J. Toughened acrylic and epoxy adhesives. In: *Synthetic Adhesives and Sealants*; Wake, W.C., Ed.; John Wiley & Sons: New York, **1987**, pp. 59-88.

[34]   Unnikrishnan, K.P.; Thachil, E.T. Toughening of epoxy resins. *Des. Monomers Polym.,* **2006**, *9*(2), 129-152.
       http://dx.doi.org/10.1163/156855506776382664

[35]   Bagheri, R.; Marouf, B.T.; Pearson, R.A. Rubber-toughened epoxies: a critical review. *Polym. Rev. (Phila. Pa.),* **2009**, *49*(3), 201-225.
       http://dx.doi.org/10.1080/15583720903048227

[36]   Ratna, D.; Banthia, A.K. Rubber toughened epoxy. *Macromol. Res.,* **2004**, *12*(1), 11-21.
       http://dx.doi.org/10.1007/BF03218989

[37]   Ge, Z.; Zhang, W.; Huang, C.; Luo, Y. Study on epoxy resin toughened by epoxidized hydroxy-terminated polybutadiene. *Materials (Basel),* **2018**, *11*(6), 932.
       http://dx.doi.org/10.3390/ma11060932 PMID: 29857534

[38]   Kishi, H.; Uesawa, K.; Matsuda, S.; Murakami, A. Adhesive strength and mechanisms of epoxy resins toughened with pre-formed thermoplastic polymer particles. *J. Adhes. Sci. Technol.,* **2005**, *19*(15), 1277-1290.
       http://dx.doi.org/10.1163/156856105774784402

[39]   Ghozali, M.; Triwulandari, E.; Haryono, A. Preparation and characterization of polyurethane modified epoxy with various types of polyol. *Macromol. Symp.,* **2015**, *353*(1), 154-160.
       http://dx.doi.org/10.1002/masy.201550321

[40]   Akimoto, K. Crack-growth resistance. *Thermoset Resin,* **1990**, *11*(4), 35-47.

[41]   Thomas, R.; Sinturel, C.; Thomas, S.; Akiaby, E. Introduction. In: *Micro- and Nanostructured Epoxy/Rubber Blends*; Thomas, S.; Sinturel, C.; Thomas, R., Eds.; Wiley-VCH Verlag: Weinheim, Germany, **2014**, p. 3.

http://dx.doi.org/10.1002/9783527666874.ch1

[42]   Paraskar, P.M.; Major, I.; Ladole, M.R.; Doke, R.B.; Patil, N.R.; Kulkarni, R.D. Dimer fatty acid – A renewable building block for high-performance polymeric materials. *Ind. Crops Prod.*, **2023**, *200*, 116817.

http://dx.doi.org/10.1016/j.indcrop.2023.116817

[43]   *Encyclopedia of chemical technology,* 5th ed; Wiley: New York, **2005**, Vol. 10, p. 394.

[44]   Mark, H.F. *Encyclopedia of chemical technology,* 4th ed; Wiley: New York, **2014**, Vol. 9, p. 727.

[45]   Guzmán, D.; Ramis, X.; Fernández-Francos, X.; Serra, A. Preparation of click thiol-ene/thiol-epoxy thermosets by controlled photo/thermal dual curing sequence. *RSC Advances*, **2015**, *5*(123), 101623-101633.

http://dx.doi.org/10.1039/C5RA22055F

[46]   *Mark, H.F. Encyclopedia of chemical technology,* 4th ed; Wiley: New York, **2014**, Vol. 9, p. 740.

[47]   Mark, H.F. *Encyclopedia of chemical technology,* 4th ed; Wiley: New York, **2014**, Vol. 9, p. 736.

[48]   *Mark, H.F. Encyclopedia of chemical technology,* 4th ed; Wiley: New York, **2014**, Vol. 9, p. 738.

[49]   Mark, H.F. *Encyclopedia of chemical technology,* 4th ed; Wiley: New York, **2014**, Vol. 9, p. 744.

[50]   Vidil, T.; Tournilhac, F.; Musso, S.; Robisson, A.; Leibler, L. Control of reactions and network structures of epoxy thermosets. *Prog. Polym. Sci.*, **2016**, *62*, 126-179.

http://dx.doi.org/10.1016/j.progpolymsci.2016.06.003

[51]   Petrie, E.M. *Epoxy Adhesive Formulations*; McGraw-Hill: New York, **2006**, pp. 85-110.

[52]   Dewprashad, B.; Eisenbraun, E.J. Fundamentals of epoxy formulation. *J. Chem. Educ.*, **1994**, *71*(4), 290-294.

http://dx.doi.org/10.1021/ed071p290

[53]   Junid, R.; Siregar, J.P.; Endot, N.A.; Razak, J.A.; Wilkinson, A.N. Optimization of glass transition temperature and pot life of epoxy blends using response surface methodology. *Polymers (Basel)*, **2021**, *13*(19), 3304.

http://dx.doi.org/10.3390/polym13193304 PMID: 34641120

[54]   Muller, B.; Rath, W. *Formulating Adhesives and Sealants*; Vincentz Network: Hanover, Germany, **2010**, pp. 137-161.

[55]   Mark, H.F. *Encyclopedia of chemical technology,* 4th ed; Wiley: New York, **2014**, Vol. 9, p. 752.

[56]   Mousavi, S.R.; Estaji, S.; Raouf Javidi, M.; Paydayesh, A.; Khonakdar, H.A.; Arjmand, M.; Rostami, E.; Jafari, S.H. Toughening of epoxy resin systems using core–shell rubber particles: a literature review. *J. Mater. Sci.*, **2021**, *56*(33), 18345-18367.

http://dx.doi.org/10.1007/s10853-021-06329-8

[57]   Baek, D.; Sim, K.B.; Kim, H.J. Mechnical characterization of core-shell rubber/epoxy polymers for automotive structural adhesives as a function of operating temperature. *Polymers (Basel),* **2021**, *13*(5), 734.

http://dx.doi.org/10.3390/polym13050734 PMID: 33673513

[58]   Chen, J.; Kinloch, A.J.; Sprenger, S.; Taylor, A.C. The mechanical properties and toughening mechanisms of an epoxy polymer modified with polysiloxane-based core-shell particles. *Polymer (Guildf.),* **2013**, *54*(16), 4276-4289.

http://dx.doi.org/10.1016/j.polymer.2013.06.009

[59]   Thitsartarn, W.; Fan, X.; Sun, Y.; Yeo, J.C.C.; Yuan, D.; He, C. Simultaneous enhancement of strength and toughness of epoxy using POSS-Rubber core–shell nanoparticles. *Compos. Sci. Technol.,* **2015**, *118*, 63-71.

http://dx.doi.org/10.1016/j.compscitech.2015.06.011

[60]   Chae, G.S.; Park, H.W.; Lee, J.H.; Shin, S. Comparative study on the impact wedge-peel performance of epoxy-based structural adhesives modified with different toughening agents. *Polymers (Basel),* **2020**, *12*(7), 1549.

http://dx.doi.org/10.3390/polym12071549 PMID: 32668731

[61]   Kenig, S.; Dodiuk, H.; Otorgust, G.; Gomid, S. Nanocomposite polymer adhesives: a critical review. *Rev. Reviews of Adhesion and Adhesives,* **2019**, *7*(2), 93-168.

http://dx.doi.org/10.7569/RAA.2019.097306

[62]   Tsang, W.L.; Taylor, A.C. Fracture and toughening mechanisms of silica- and core–shell rubber-toughened epoxy at ambient and low temperature. *J. Mater. Sci.,* **2019**, *54*(22), 13938-13958.

http://dx.doi.org/10.1007/s10853-019-03893-y

[63]   International Organization for Standardization. *Plastics—Epoxy—Test Method*; ISO Standard 18280:2010, **2010.**

[64]   ASTM International. *Standard Guide for Testing Epoxy Resins*; ASTM D 4142-89, **2009**.

[65]   ASTM International. *Standard Specification for Epoxy Adhesive for Bonding Metallic and Nonmetallic Materials*; ASTM D6412-12 (D6412), M-99, **2012.**

[66]   Biron, M. *Thermosets and Composites,* 2nd ed,; Elsevier: London, UK, **2014**, pp. 211-212.

[67]   Henkel Corporation. *Technical Data Sheet of Loctite EA E-00CL.* https://datasheets.tdx.henkel.com/LOCTITE-EA-E-00CL-en_GL.pdf (accessed December 20, 2023).

[68]   Henkel Corporation. *Technical Data Sheet of Loctite EA E-30CL.* https://datasheets.tdx.henkel.com/LOCTITE-EA-E-30CL-en_GL.pdf (accessed December 20, 2023).

[69]   Henkel Corporation. *Technical Data Sheet of Loctite EA E-60HP.* https://datasheets.tdx.henkel.com/LOCTITE-EA-E-60HP-en_GL.pdf (accessed December 20, 2023).

[70]   Chen, C.; Iwasaki, S.; Kanari, M.; Lu, D. High Temperature Resistant Resin Composition. *Patent Application WO2022/117013,* **2022.**

[71]   Güthner, T.; Hammer, B. Curing of epoxy resins with dicyandiamide and urones. *J. Appl. Polym. Sci.,* **1993**, *50*(8), 1453-1459.
http://dx.doi.org/10.1002/app.1993.070500817

[72]   Petrie, E.M. *Epoxy Adhesive Formulations*; McGraw-Hill: New York, **2006**, p. 250.

[73]   Voytekunas, V.Y.; Ng, F.L.; Abadie, M.J.M. Kinetics study of the UV-initiated cationic polymerization of cycloaliphatic diepoxide resins. *Eur. Polym. J.,* **2008**, *44*(11), 3640-3649.
http://dx.doi.org/10.1016/j.eurpolymj.2008.08.043

[74]   Golaz, B.; Michaud, V.; Leterrier, Y.; Månson, J.A.E. UV intensity, temperature and dark-curing effects in cationic photo-polymerization of a cycloaliphatic epoxy resin. *Polymer (Guildf.),* **2012**, *53*(10), 2038-2048.
http://dx.doi.org/10.1016/j.polymer.2012.03.025

[75]   Chen, C.; Li, B.; Wang, C.; Iwasaki, S.; Kanari, M.; Lu, D. UV and thermal cure epoxy adhesive. In: *Painting and Coatings Industry*; Yalmaz, F., Ed.; IntechOpen: London, UK, **2019**, pp. 71-85.
http://dx.doi.org/10.5772/intechopen.82168

[76]   Gan, Y.; Chen, C.; Terada, K. Cationically photocurable epoxy resin composition. U.S. Patent 7456230, **2008**.

[77]   Velankar, S.; Pazos, J.; Cooper, S.L. High-performance UV-curable urethane acrylates *via* deblocking chemistry. *J. Appl. Polym. Sci.,* **1996**, *62*(9), 1361-1376.
http://dx.doi.org/10.1002/(SICI)1097-4628(19961128)62:9<1361::AID-APP6>3.0.CO;2-F

[78]   Fourassier, J.; Lalevee, J. *Photoinitiator for Polymer Synthesis*; Wiley-VCH Verlag: Weinheim, Germany, **2012**, pp. 41-72.
http://dx.doi.org/10.1002/9783527648245.ch4

[79]   Chen, C. *One-Component Instant Bonding Epoxy Adhesives*; in: *Proceedings of IUPAC – MACRO 2020+ The 48th World Polymer Congress,* 2OS9-2, Jeju, Korea, May 16 – 20, **2021**.

[80]   Bitzer, K.; By, A. Active alignment for cameras in mobile devices and automotive applications. *2010 12th Electronics Packaging Technology Conference,* Singapore, **2010**. pp. 260-264.

[81]   Arimitsu, K.; Sugioka, S.; Watanabe, K.; Furutani, M. Anionic UV curing of epoxy resins containing dispersed scaly silica modified with base-amplifying groups. *Prog. Org. Coat.,* **2017**, *113*, 185-188.
http://dx.doi.org/10.1016/j.porgcoat.2017.09.006

[82]   Chen, L.; Zheng, Y.; Meng, X.; Wei, G.; Dietliker, K.; Li, Z. Delayed thiol-epoxy photopolymerization: a general and effective strategy to prepare thick composites. *ACS Omega,* **2020**, *5*(25), 15192-15201.
http://dx.doi.org/10.1021/acsomega.0c01170 PMID: 32637792

[83]   Fukui, H.; Kondo, S.; Arimitsu, K. Photocurable composition. *U.S. Patent 8536242,* **2013**.

[84]   Chen, C. One component composition based on epoxy resin. *Eur. Patent Appl. 3916033*, **2021**.

[85]   Kumar, S.; Samal, S.K.; Mohanty, S.; Nayak, S.K. Recent development of biobased epoxy resin: a review. *Polym. Plast. Technol. Eng.,* **2018**, *57*(3), 133-155.
http://dx.doi.org/10.1080/03602559.2016.1253742

[86]   Cailoll, S.; Boutevin, B.; Pascault, J. Bio-sourced epoxy monomer and polymers. In: *Handbook of Adhesive Technology,* 3rd ed; Pizzi, A.; Mittal, K.L., Eds.; CRC Press: Boca Raton, FL, **2018**, pp. 443-470.

[87]   Heinrich, L.A. Future opportunities for bio-based adhesives – advantages beyond renewability. *Green Chem.,* **2019**, *21*(8), 1866-1888.
http://dx.doi.org/10.1039/C8GC03746A

[88]   Mashouf Roudsari, G.; Mohanty, A.K.; Misra, M. Green approaches to engineer tough biobased epoxies: a review. *ACS Sustain. Chem.& Eng.,* **2017**, *5*(11), 9528-9541.
http://dx.doi.org/10.1021/acssuschemeng.7b01422

[89]   Hemmilä, V.; Adamopoulos, S.; Karlsson, O.; Kumar, A. Development of sustainable bio-adhesives for engineered wood panels – A Review. *RSC Advances,* **2017**, *7*(61), 38604-38630.
http://dx.doi.org/10.1039/C7RA06598A

[90]   Hase, E.; Kitano, M. The role and place of risk assessment in chemical management especially in Japan, Europe and U.S.A. *Japanese J. Risk Anal.,* **2012**, *22*, 63-72.

[91]   Packham, D.E. The environmental impact of adhesives. In: *Eco-Efficient Construction and Building Materials Life Cycle Assessment (LCA), Eco-Labelling and Case Studies*; Pacheco-Torgal, F.; Cabeza, L.F.; Labrincha, J.; Magalhiies, A., Eds.; Woodhead Publishing: Cambridge, UK, **2014**, pp. 338-367.
http://dx.doi.org/10.1533/9780857097729.2.338

[92]   Rijk, R.; Veraart, R. *Global Legislation for Food Packaging Materials*; Wiley – VCH: Weinheim, Germany, **2010**, pp. 13-15.
http://dx.doi.org/10.1002/9783527630059

[93]   Santosh, E.; Yadav, K.; Palmese, G. R.; Stanzione, J. F. Recent advances in bio-based epoxy resins and bio-based epoxy curing agents. *J. Appl. Polym. Sci.* **2016**, 44103.

[94]   Auvergne, R.; Caillol, S.; David, G.; Boutevin, B.; Pascault, J.P. Biobased thermosetting epoxy: present and future. *Chem. Rev.,* **2014**, *114*(2), 1082-1115.
http://dx.doi.org/10.1021/cr3001274 PMID: 24125074

[95]   Benega, M.A.G.; Raja, R.; Blake, J.I.R. A preliminary evaluation of bio-based epoxy resin hardeners for maritime application. *Procedia Eng.,* **2017**, *200*, 186-192.
http://dx.doi.org/10.1016/j.proeng.2017.07.027

[96]   Hambleton, K.M.; Stanzione, J.F., III Synthesis and characterization of a low-molecular-weight novolac epoxy derived from lignin-inspired phenolics. *ACS Omega,* **2021**, *6*(37), 23855-23861.
http://dx.doi.org/10.1021/acsomega.1c02799 PMID: 34568665

[97]   Wang, Z.; Gnanasekar, P.; Nair, S.S.; Yi, S.; Yan, N. Curing behavior and thermomechanical performance of bio-epoxy resin synthesized from vanillyl alcohol: effects of the curing agent. *Polymers (Basel),* **2021,** *13*(17), 2891.

http://dx.doi.org/10.3390/polym13172891 PMID: 34502931

[98]   Nikafshar, S.; Wang, J.; Dunne, K.; Sangthonganotai, P.; Nejad, M. Choosing the right lignin to fully replace bisphenol A in epoxy resin. *ChemSusChem,* **2021,** *14*(4), 1184-1195.

http://dx.doi.org/10.1002/cssc.202002729 PMID: 33464727

[99]   Zhang, Z.; Li, J.; Zhang, Y.; Jin, Z. Kinetics of partially depolymerized lignin as co-curing agent for epoxy resin. *Int. J. Biol. Macromol.,* **2020,** *150,* 786-792.

http://dx.doi.org/10.1016/j.ijbiomac.2020.02.059 PMID: 32061695

[100]   Aziz, T.; Fan, H.; Zhang, X.; Khan, F.U.; Fahad, S.; Ullah, A. Adhesive properties of bio-based epoxy resin reinforced by cellulose nanocrystal additives. *Journal of Polymer Engineering,* **2020,** *40*(4), 314-320.

http://dx.doi.org/10.1515/polyeng-2019-0255

[101]   Ruiz, Q.; Pourchet, S.; Placet, V.; Plasseraud, L.; Boni, G. New eco-friendly synthesized thermosets from isoeugenol-based epoxy resin. *Polymers (Basel),* **2020,** *12*(1), 229.

http://dx.doi.org/10.3390/polym12010229 PMID: 31963401

[102]   Yang, X.; Wang, C.; Li, S.; Huang, K.; Li, M.; Mao, W.; Cao, S.; Xia, J. Study on the synthesis of bio-based epoxy curing agent derived from myrcene and castor oil and the properties of the cured products. *RSC Advances,* **2017,** *7*(1), 238-247.

http://dx.doi.org/10.1039/C6RA24818G

[103]   Merighi, S.; Mazzocchetti, L.; Benelli, T.; Giorgini, L. Evaluation of novel bio-based amino curing agent systems for epoxy resins: effect of tryptophan and guanine. *Processes (Basel),* **2020,** *9*(1), 42.

http://dx.doi.org/10.3390/pr9010042

[104]   Ding, C.; Matharu, A.S. Recent developments on biobased curing agents: a review of their preparation and use. *ACS Sustain. Chem.& Eng.,* **2014,** *2*(10), 2217-2236.

http://dx.doi.org/10.1021/sc500478f

[105]   Song, X.; Deng, Z.P.; Li, C.B.; Song, F.; Wang, X.L.; Chen, L.; Guo, D.M.; Wang, Y.Z. A bio-based epoxy resin derived from p-hydroxycinnamic acid with high mechanical properties and flame retardancy. *Chin. Chem. Lett.,* **2022,** *33*(11), 4912-4917.

http://dx.doi.org/10.1016/j.cclet.2021.12.067

[106]   Gonçalves, F.A.M.M.; Ferreira, P.; Alves, P. Synthesis and characterization of itaconic-based epoxy resin: Chemical and thermal properties of partially biobased epoxy resins. *Polymer (Guildf.),* **2021,** *235,* 124285.

http://dx.doi.org/10.1016/j.polymer.2021.124285

[107]   Chong, K.L.; Lai, J.C.; Rahman, R.A.; Adrus, N.; Al-Saffar, Z.H.; Hassan, A.; Lim, T.H.; Wahit, M.U. A review on recent approaches to sustainable bio-based epoxy vitrimer from epoxidized vegetable oils. *Ind. Crops Prod.,* **2022,** *189,* 115857.

http://dx.doi.org/10.1016/j.indcrop.2022.115857

[108]   Kalita, D.J.; Tarnavchyk, I.; Kalita, H.; Chisholm, B.J.; Webster, D.C. Novel bio-based epoxy resins from eugenol derived copolymers as an alternative to DGEBA resin. *Prog. Org. Coat.,* **2023**, *178*, 107471.

http://dx.doi.org/10.1016/j.porgcoat.2023.107471

[109]   Ma, Y.; Kou, Z.; Hu, Y.; Zhou, J.; Bei, Y.; Hu, L.; Huang, Q.; Jia, P.; Zhou, Y. Research advances in bio-based adhesives. *Int. J. Adhes. Adhes.,* **2023**, *126*, 103444.

http://dx.doi.org/10.1016/j.ijadhadh.2023.103444

[110]   Gaina, C.; Ursache, O.; Gaina, V.; Serban, A.M.; Asandulesa, M. Novel bio-based materials: from castor oil to epoxy resins for engineering applications. *Materials (Basel),* **2023**, *16*(16), 5649.

http://dx.doi.org/10.3390/ma16165649 PMID: 37629941

[111]   Pappa, C.P.; Torofias, S.; Triantafyllidis, K.S. Sub-micro organosolv lignin as bio-based epoxy polymer component: a sustainable curing agent and additive. *ChemSusChem,* **2023**, *16*(13), e202300076.

http://dx.doi.org/10.1002/cssc.202300076 PMID: 36912587

CHAPTER 2

UV Cure Cationic Epoxy Technology

Abstract: UV cure cationic epoxy adhesives are composed of epoxy resins and cationic photoinitiators with additives and modifiers. Cycloaliphatic epoxy resins are the main epoxy resins used for cationic epoxy formulations. Oxetanes are often combined with epoxy resins to improve curability and adjust viscosity. Cationic photoinitiators are all onium salts, composed of an organic cation with an inorganic anion. UV cure cationic epoxy adhesives can be cured quickly *via* UV light radiation and have been very successful in electronics assembly and general bonding applications. UV cure equipment, formulating, testing, and evaluation of UV cure cationic epoxy adhesives are described.

Keywords: Cationic photoinitiator, LED-UV lamp, Mercury lamp, Metal halide lamp, Oxetane, Sensitizer, UV cure equipment, UV cure, UV radiometer.

UV CURE CATIONIC EPOXY CHEMISTRY

Ultra-violet light (UV) cure cationic epoxy adhesives can be cured quickly and have been very successful in electronics assembly and general bonding applications, such as general glass bonding, optical parts assembly, image sensor module assembly, display panel bonding, and module assembly, that require fast production speed and high adhesion performance. UV cationic epoxy adhesives are primarily composed of epoxy resins, reactive additives, and cationic photoinitiators with additives and modifiers [1-10]. UV cure cationic epoxy adhesives provide significant advantages over UV cure acrylics.

✓ No oxygen inhibition.
✓ Low cure shrinkage.
✓ Good adhesion and chemical resistance.
✓ Dark curability.

Epoxy Resins and Oxetanes

Cycloaliphatic type epoxy resins are the main epoxy resins used for cationic epoxy adhesives because they can cure faster *via* cationic polymerization than normal glycidyl ether type epoxy resins such as bisphenol A epoxy resin. (3',4'-Epoxycyclohexane) methyl-3,4-epoxycyclohexylcarboxylate, often called ECC, is the principal cycloaliphatic epoxy resin used, providing a high cross-linked rigid chemical structure with high transition temperature after full cure. 3,4-Epoxycyclohexylmethyl-3',4'-epoxycyclohexanecarboxylate-modified epsilon - caprolactone, or bis(3,4-epoxycyclohexyl) adipate, is often combined to improve flexibility and toughness of ECC.

Hydrogenated bisphenol A epoxy resin can be prepared by the hydrogenation of bisphenol A epoxy resin, as shown in Fig. (**1**) [11]. Hydrogenated bisphenol A epoxy resin can cure faster *via* cationic polymerization and shows better optical properties and light resistance than bisphenol A epoxy resin due to no existence of aromatic structure. Hydrogenated bisphenol A epoxy resin is a good alternative for cycloaliphatic epoxy resins where higher adhesion and humidity reliability performance are required.

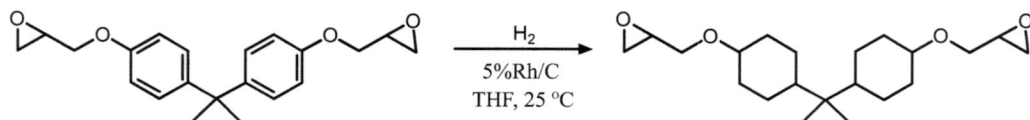

Fig. (1). Synthesis of hydrogenated bisphenol A epoxy resin.

Oxetanes are heterocyclic organic compounds containing the oxetane group, having a four-membered ring with three carbon atoms and one oxygen atom. Oxetanes cure faster than epoxy resins *via* cationic polymerization. 3,3'-[Oxybis(methylene)] bis(3-ethyloxetane), or dioxetanyl ether (abbreviated as DOX), is a difunctional oxetane with very low viscosity. 3-Ethyl-3-hydroxylmethyl-oxetane, or oxetane alcohol (abbreviated as OXE), is an oxetane having a hydroxy group. Similar to polyol, the hydroxy group of OXE can enhance the cure speed of cationic polymerization through chain transfer. Oxetanes are often combined with epoxy resins to improve curability and adjust viscosity.

Chemical structure, viscosity, CAS No., and key features of cycloaliphatic epoxy resins, hydrogenated bisphenol A epoxy resin, and oxetanes are summarized in Fig. (**2**).

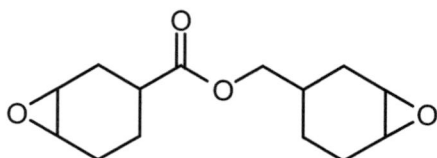

Viscosity: 300 mPa.s/25°C
CAS. No.: 2386-87-0
EEW: 135
High Tg

(3',4'-Epoxycyclohexane)methyl-3,4-epoxycyclohexylcarboxylate (ECC)

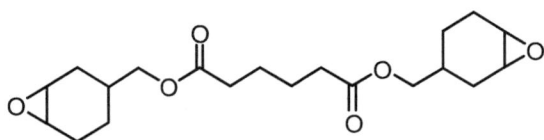

Viscosity: 600 mPa.s/25°C
CAS No.: 139198-19-9
EEW: 200
Flexibility

3,4-Epoxycyclohexylmethyl-3',4'-epoxycyclohexanecarboxylate-modified epsilon-caprolactone

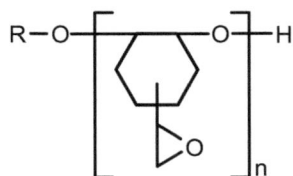

Softening point: 75°C
CAS No.: 244772-00-7
EEW: 177
High thermal resistance

Poly [(2-ethylene epoxide) - 1,2-cyclohexanediol] 2-ethyl-2 - (hydroxymethyl) - 1,3-propylene glycol ether

Softening point: 185 - 189°C
CAS No.: 81-21-0
EEW: 82 -85
High thermal resistance

Dicyclopentadienediepoxide

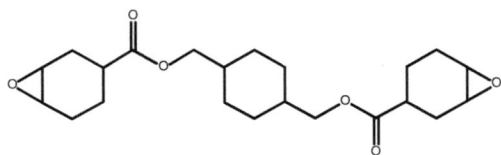

Softening point: 60°C
CAS No.: 20249-12-1
EEW: 190 - 210
Good weatherability

1,4-Cyclohexanedimethanol bis(3,4-epoxycyclohexanecarboxylate)

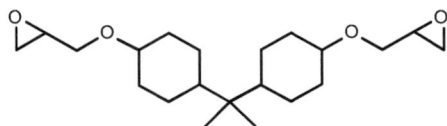

Viscosity: 2000 mPa.s/25°C
CAS No.: 30583-72-3
EEW: 220
Good adhesion

Hydrogenated bisphenol A epoxy resin

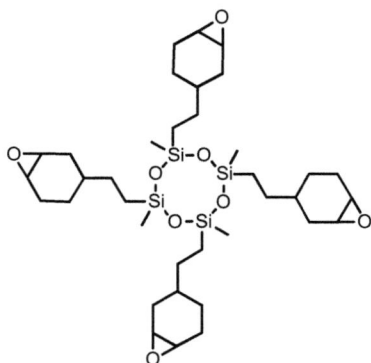

Viscosity: 3000 mPa.s/25°C
CAS No.: 121225-98-7
Molar mass: 200
Fast curability, high Tg

Polymethyl beta-(3, 4-epoxycyclohexyl) ethyl siloxane

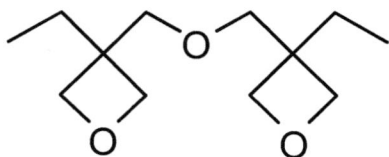

Viscosity: 13 mPa.s/25°C
CAS No.: 18934-00-4
Molar mass: 214
Fast curability

3,3'-[Oxybis(methylene)]bis(3-ethyloxetane) (DOX)

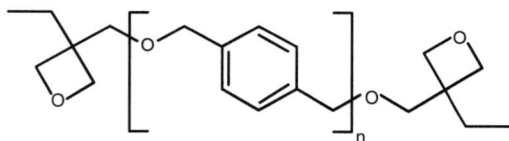

Viscosity: 13 mPa.s/25°C
CAS No.: 142627-97-2
Molar mass: 334
Fast curability

1,4-bis[(3-ethyl-3-oxetanylmethoxy)methyl]benzene

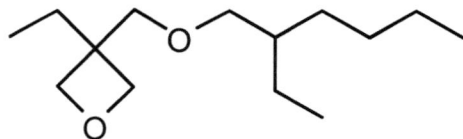

Viscosity: 3 - 6 mPa.s/25°C
CAS No.: 298695-60-0
Molar mass: 228
Fast curability, high diluency

3-Ethyl-3-(2-ethylhexoxymethyl)oxetane

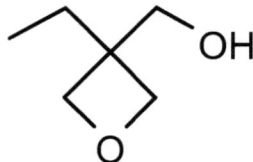

Viscosity: 20 mPa.s/25°C
CAS No.: 3047-32-3
Molar mass: 116
Fast curability, diluency

3-Ethyl-3-hydroxylmethyl-oxetane (OXE)

Fig. (2). Typical epoxy resins and oxetanes used in cationic epoxy adhesives.

Cationic Photoinitiators

Cationic photoinitiators commercialized are all onium salts due to their balanced photosensitivity, cationic reactivity, and latency. Onium salts are ionic compounds composed of an organic cation with an inorganic anion. The organic cation determines photosensitivity and light absorption efficiency during UV light radiation, while the inorganic anion determines the strength of the Bronsted or Lewis acid generated during the photodecomposition process. In general, the strength of the acid generated depends on the nucleophilicity, and its reactivity decreases in the sequence of:

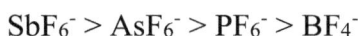

$$SbF_6^- > AsF_6^- > PF_6^- > BF_4^-$$

Fig. (3) shows photodecomposition and the Bronsted or Lewis acid formation mechanism of onium salt [12].

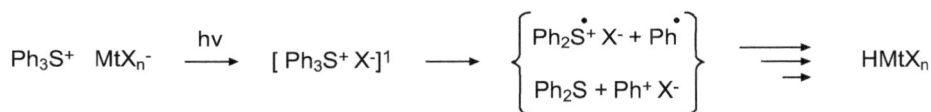

Fig. (3). Photodecomposition scheme of onium salt.

Sulfonium salts combined with hexafluoroantimonate SbF_6^- anion are the most common cationic photoinitiators used so far. Fig. (4) shows the formation mechanism of bronsted acid of a trisulfonium salt type photoininitiator *via* heterolytic and homolytic cleavage routes [13].

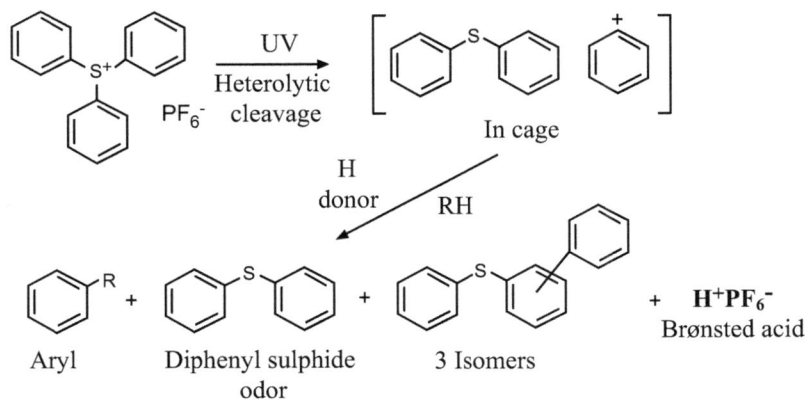

(a)

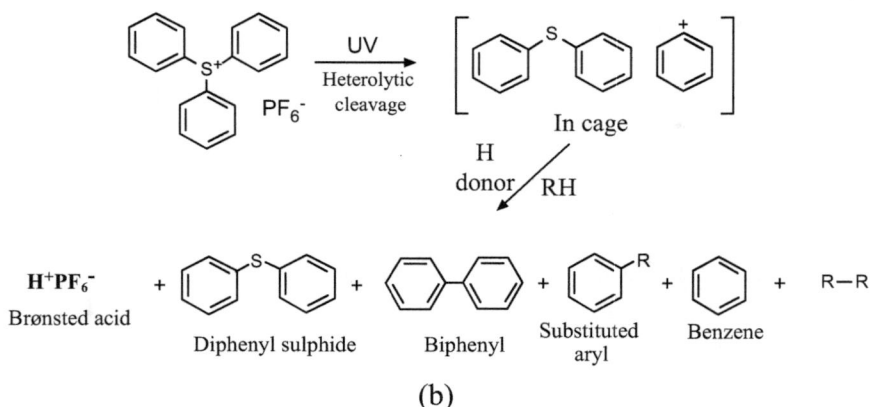

(b)

Fig. (4). Bronsted acid formation of sulfonium salt *via* heterolytic (a), and homolytic cleavage (b).

Fig. (5) shows the chemical structure and key properties of the most commonly used cationic photoinitiator. It is a mixture of sulfonium salts with the hexafluoroantimonate SbF_6^- anion, designed for more efficient UV light absorption. Commercial cationic photoinitiator products are typically supplied in liquid solution by dissolving the onium salt in propylene carbonate, a high boiling point solvent, for easy handling. Fig. (6) shows the absorption spectrum of Speedcure 976 supplied by Arkema Corporation [14]. Onium salts with hexafluorophosphate PF_6^- anion have been increasingly used recently due to toxic concerns on the antimony compounds.

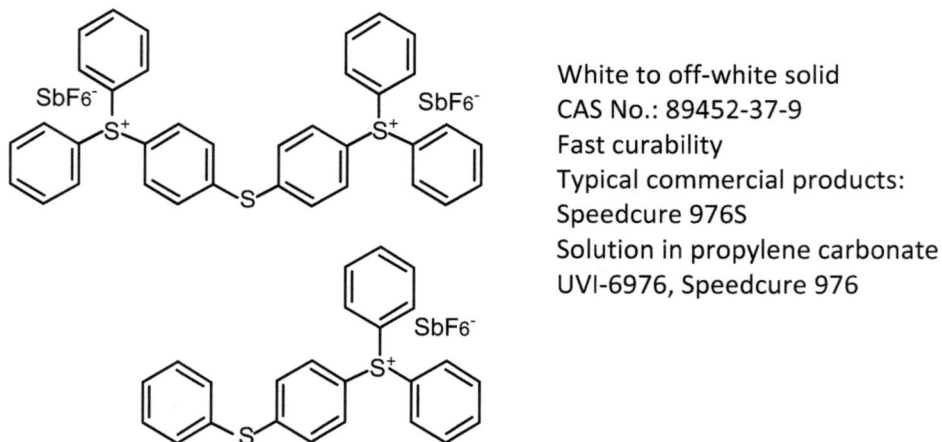

White to off-white solid
CAS No.: 89452-37-9
Fast curability
Typical commercial products:
Speedcure 976S
Solution in propylene carbonate
UVI-6976, Speedcure 976

Sulfanediyldibenzene-4,1-diyl) bis(diphenylsulfonium) bis(hexafluoroantimonate)

Fig. (5). Chemical structure and key properties of typical sulfonium salt with SbF_6^-

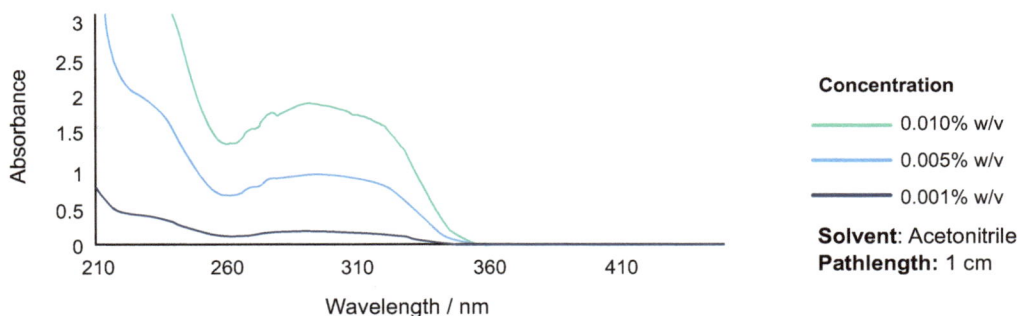

Fig. (6). Absorption spectrum of Speedcure 976.

Fig. (**7**) shows the chemical structure and key properties of another commonly used photoinitiator. It is also a mixture of sulfonium salts with the hexafluorophosphate PF_6^- anion. Commercial products are typically supplied in liquid solution by dissolving the onium salt in propylene carbonate, a high boiling point solvent, for easy handling. Fig. (**8**) shows the absorption spectrum of Speedcure 992 supplied by Arkema Corporation. The hexafluorophosphate PF_6^- anion has been increasingly used recently due to toxic concerns on the antimony compounds.

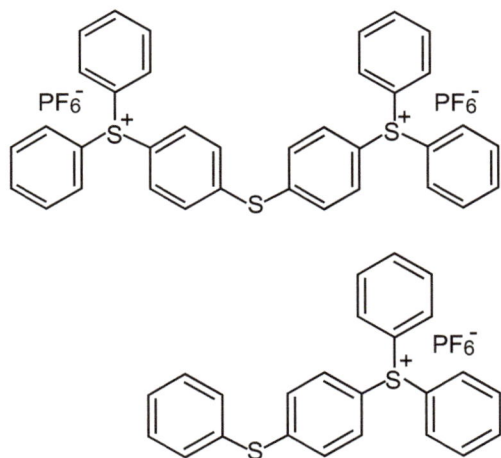

Clear yellow to amber liquid
CAS No.: 74227-35-3, 108-32-7
Good curability
Typical commercial products:
 (solution in propylene carbonate)
UVI-6992, Speedcure 992

(4-{[4-(diphenylsulfanylium)   phenyl]sulfanyl}phenyl)   diphenylsulfonium
bishexafluorophosphate

Fig. (7). Chemical structure and key properties of typical sulfonium salt with PF_6^-

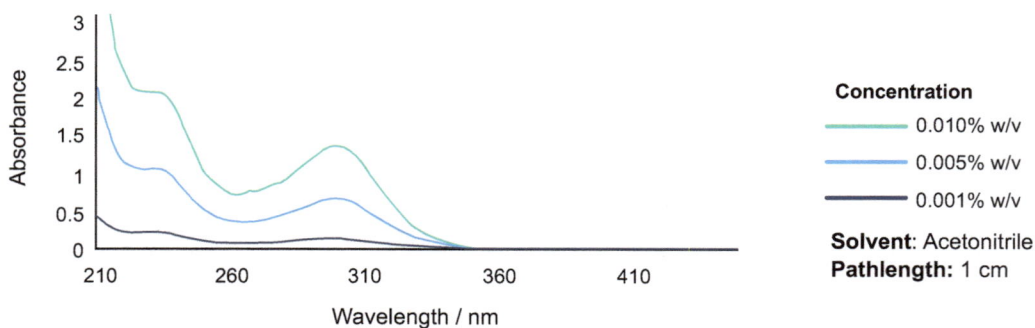

Fig. (8). Absorption spectrum of Speedcure 992.

As can be seen from Fig. (4), there is a potentially trace amount of benzene generation from commonly used sulfonium salt type cationic photoinitiator during UV radiation. There are safety concerns about benzene for some critical applications. The benzene-free type has been developed and commercialized. Fig. (9) shows the chemical structure of a benzene-free type sulfonium salt with the hexafluoroantimonate SbF_6^- anion, SP-170, supplied by ADEKA Corporation. Fig. (10) shows the chemical structure of a benzene-free type sulfonium salt with the the hexafluorphosphate PF_6^- anion, SP-150, supplied by ADEKA Corporation.

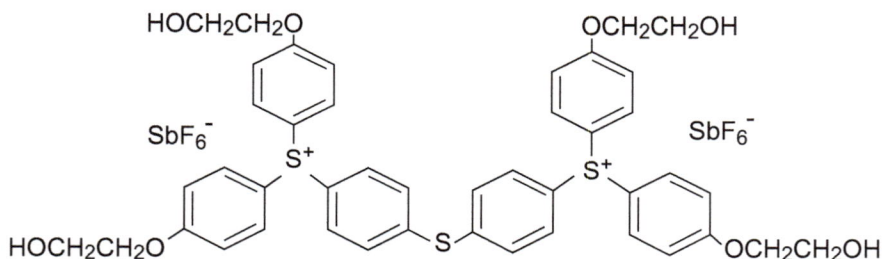

Fig. (9). Chemical structure of a benzene-free type sulfonium salt with SbF_6^-

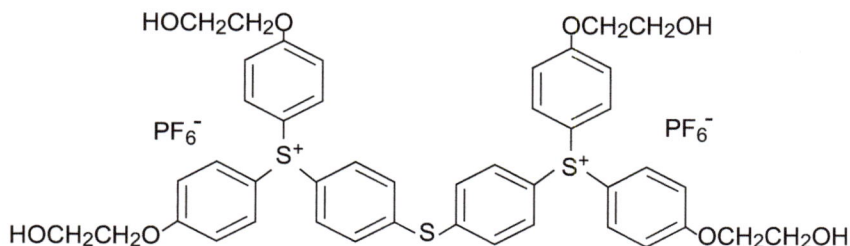

Fig. (10). Chemical structure of a benzene-free type sulfonium salt with PF_6^-

Iodonium salt types are also commonly used cationic photoinitiators. Fig. (**11**) shows the formation mechanism of bronsted acid of a diphenyliodonium salt combined with the hexafluorphosphate PF_6^- anion *via* heterolytic and homolytic cleavage routes [15]. Iodonium salts are less reactive than sulfonium salts due to shorter wavelength absorption below 300nm and are unable to uptake the 313 nm output line of conventional mercury lamps. However, its curability can be much improved with a combination of photosensitizers.

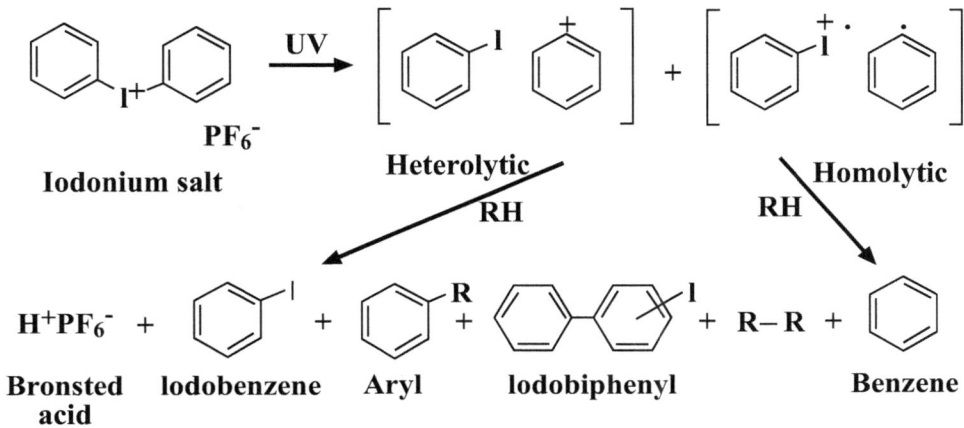

Fig. (11). Bronsted acid formation of iodonium salt *via* heterolytic and homolytic cleavage.

Fig. (**12**) shows the chemical structure and key properties of an iodonium salt with the hexafluoroantimonate SbF_6^- anion. Fig. (**13**) shows the absorption spectrum of Speedcure 937 supplied by Arkema Corporation [16]. As can be seen from the UV spectrum, its absorption is below 300 nm.

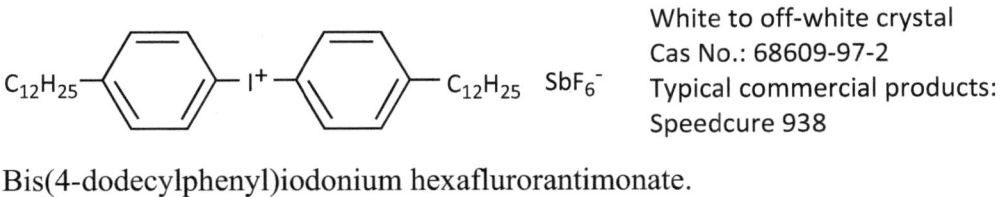

White to off-white crystal
Cas No.: 68609-97-2
Typical commercial products:
Speedcure 938

Bis(4-dodecylphenyl)iodonium hexaflurorantimonate.

Fig. (12). Chemical structure and key properties of typical iodonium salt with SbF_6^-

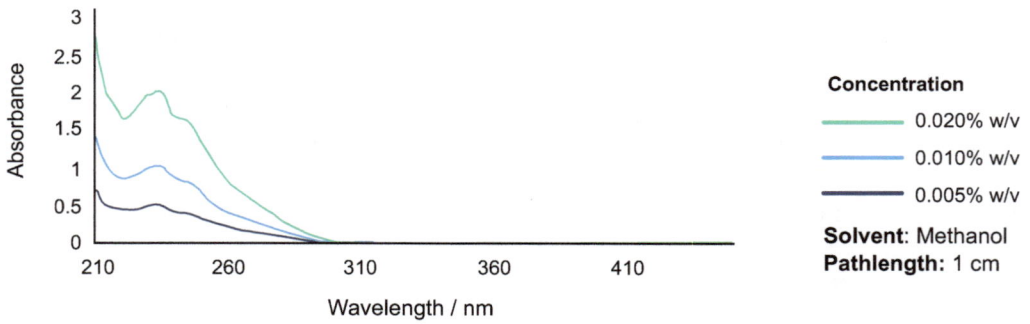

Fig. (13). UV spectrum of Speedcure 937.

Fig. (**14**) shows the chemical structure and key properties of an iodonium salt with the hexafluorophosphate PF6- anion. Fig. (**15**) shows the absorption spectrum of Speedcure 938 supplied by Arkema Corporation [17]. As can be seen from the UV spectrum, its absorption is also below 300 nm.

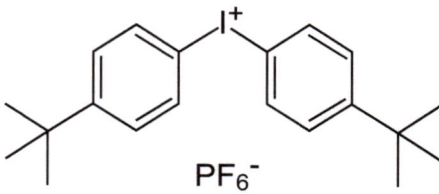

White solid
Cas No.: 61358-25-6
Typical commercial products:
Speedcure 938

Bis-(4-t-butylphenyl)-iodonium hexafluorophosphate

Fig. (14). Chemical structure and key properties of typical iodonium salt with PF_6^-

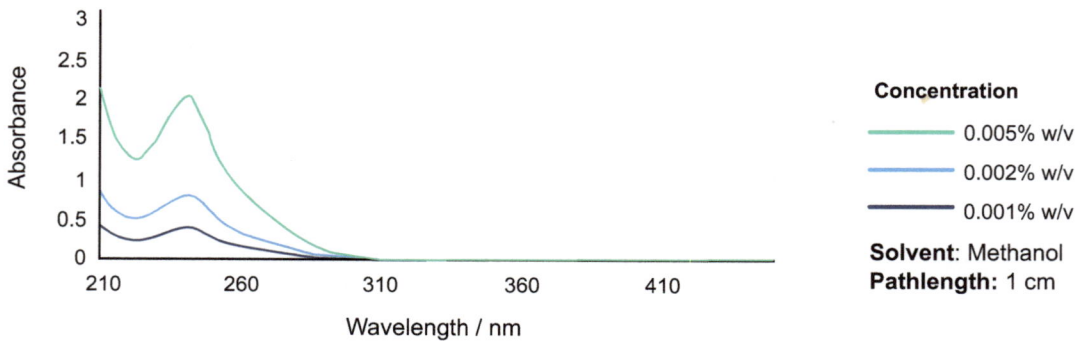

Fig. (15). UV spectrum of Speedcure 938.

Sensitizers

Most cationic photoinitiators, especially iodonium salts, absorb short wavelength ranges of UV and are generally not useful for long wavelength radiation conditions or highly pigmented cases. Sensitizers can be combined in these situations. As shown in Fig. (**16**) [18], sensitizer-absorbed UV light becomes an activated sensitizer that can react with iodonium salt *via* electron transfer to generate the acid, which will initiate cationic polymerization. Thioxanthones, such as isopropylthioxanthone (ITX), 1-chloro-4-propoxythioxanthone (CPTX), and dialkoxy anthracene derivatives such as 2-ethyl-9,10-dimethoxyanthrancene and 9,10-diethoxy anthracene are typically used as photosensitizer for iodonium salt type cationic photoinitiators. Sensitization for sulfonium salt-type cationic photoinitiators is generally not effective. Fig. (**17**) shows the chemical structure of these cationic photosensitizers. Typical sensitizers, their peak absorption wavelength, and sensitizing effects on sulfonium salt and iodonium salt are summarized in Table **1**.

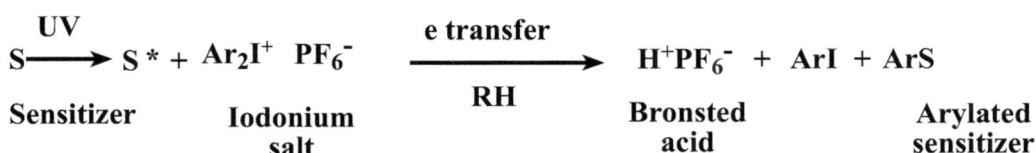

$$S \xrightarrow{\text{UV}} S* + Ar_2I^+ \ PF_6^- \xrightarrow[\text{RH}]{\text{e transfer}} H^+PF_6^- \ + \ ArI \ + \ ArS$$

Sensitizer　　　　　**Iodonium**　　　　　　　　　　**Bronsted**　　　　　　**Arylated**
　　　　　　　　　　　　salt　　　　　　　　　　　　　**acid**　　　　　　　**sensitizer**

Fig. (16). Sensitization mechanism for UV cationic iodonium salt photoinitiator.

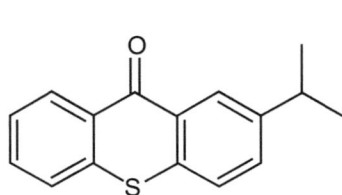

Isopropylthioxanthone (ITX)　　　　1-chloro-4-propoxythioxanthone (CPTX)

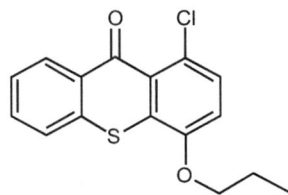

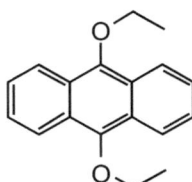

9,10-diethoxy anthracene (DEA)　　　　9,10-dibutoxy anthracene (DBA)

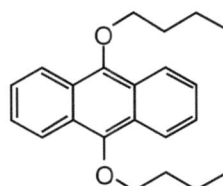

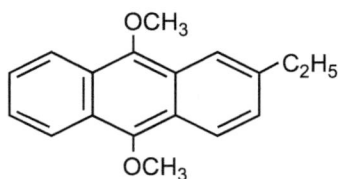

2-ethyl-9,10-dimethoxyanthrancene (EDMA)   benzophenone

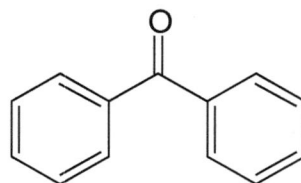

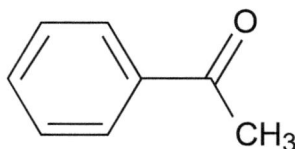

Acetophenone

Fig. (17). Chemical structure of typical cationic photosensitizers.

Table 1. Cationic photosensitizers.

Sensitizer	Absorption, nm	Sulphonium Salt	Iodonium Salt
ITX	383	No	Yes
CPTX	387	No	Yes
DEA	402	Yes	Yes
DBA	403	Yes	Yes
EDMA	401	Yes	Yes
Benzophenone	254	No	Yes
Acetophenone	278	Yes	Yes

Polymerization Mechanism

As illustrated in Fig. (**18**), the Bronsted or Lewis acid formed by photodecomposition of onium salt will react with an epoxy group to become a carbonium cation that can initiate polymerization of epoxy resin. All these reactions that start from the Bronsted or Lewis acid are thermally driven and independent of the UV source. The Bronsted or Lewis acid is very stable and will continue to initiate cationic polymerization after the UV source is moved out. This cationic reaction phenomenon is often called 'dark reaction' of UV cure cationic epoxies.

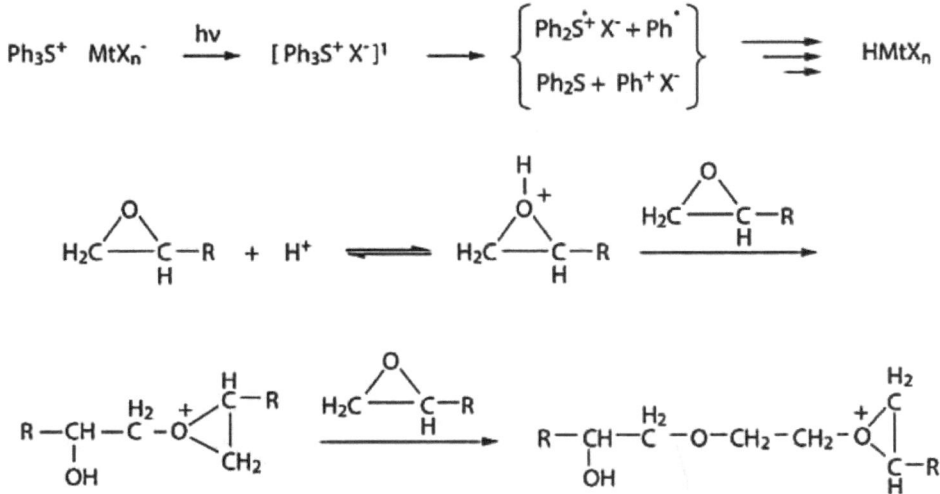

Fig. (18). UV cationic polymerization mechanism of epoxy adhesives.

UV CURE EQUIPMENT

UV Lamp

Ultraviolet and near visible light with wavelength between 200 nm and 450 nm is used for light curing applications. The spectrum range is further defined as UV-Vis (400 – 450 nm), UV A (320 – 400 nm), UV B (280 – 320 nm), and UV C (200 – 280 nm). Short-wavelength light has a higher energy level but is less penetrative than longer-wavelength light. In contrast, longer wavelength light has deeper penetrative but lower energy levels than shorter wavelength light. UV A is the most important light range for UV curing, while UV C light works more effectively for surface cure. The electromagnet spectrum and the UV-visible light spectrum are shown in Fig. (**19**) [19]. The wavelength ranges and their main uses for UV cure applications are summarized in Table **2**.

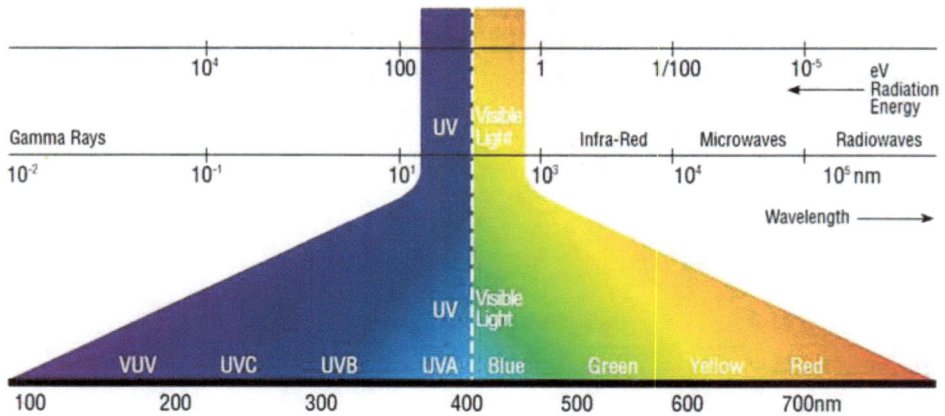

Fig. (19). The electromagnetic spectrum and the UV-visible light spectrum.

Table 2. Wavelength range for UV cure applications.

Term	Wavelength (nm)	Main usages
UVA	315 – 400	UV polymerization; fluorescent inspection
UVB	280 – 315	UV polymerization; suntanning
UVC	200 – 280	Surface UV cure
UV-Vis	400 – 480	Depth cure

Conventional UV lamps are the medium-pressure mercury lamp (MPM) and the doped metal halide lamp. As shown in Fig. (**20**) [20], the standard mercury and metal halide lamp is composed of a sealed quartz tube and electrodes in each end, with a certain amount of mercury or metal-doped mercury and a small amount of a starter gas, usually argon sealed in the tube. When the lamp is switched on, the low-pressure argon is ionized, which allows current to flow and heats the lamp to an operating temperature of around 600 – 800°C. Mercury in the tube is then vaporized, and the mercury vapor becomes exited *via* an energy source, generating a wide range of emission lines throughout the UV and visible light spectrum. When the lamp is switched off, the mercury ions and electrons combine to form back to mercury.

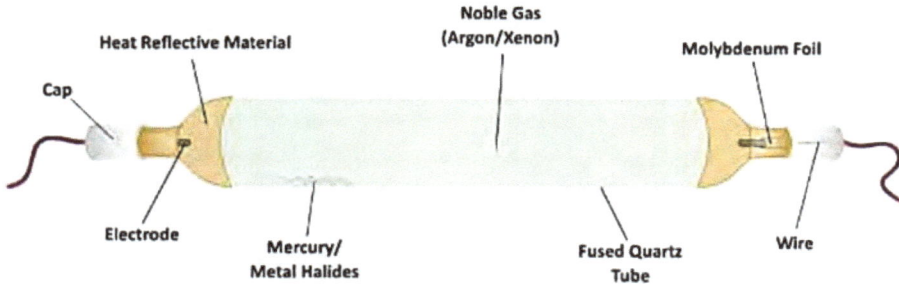

Fig. (20). Structure of the standard mercury and metal halide lamp.

As shown in Fig. (**21**), the medium-pressure mercury lamp provides a wide range of UV and UV-vis spectra. Its strong UV A output line at 366 nm plays a key role in UV curing. Its UV-vis output line at 404 nm and 436 nm is more penetrative and helps in better depth cure. Its UV C output line at 254 nm is effective for surface cure, while its UV B output line at 313 nm helps in both surface cure and depth cure. The visible output lines at 550 nm and 580 nm are usually of no use in UV curing.

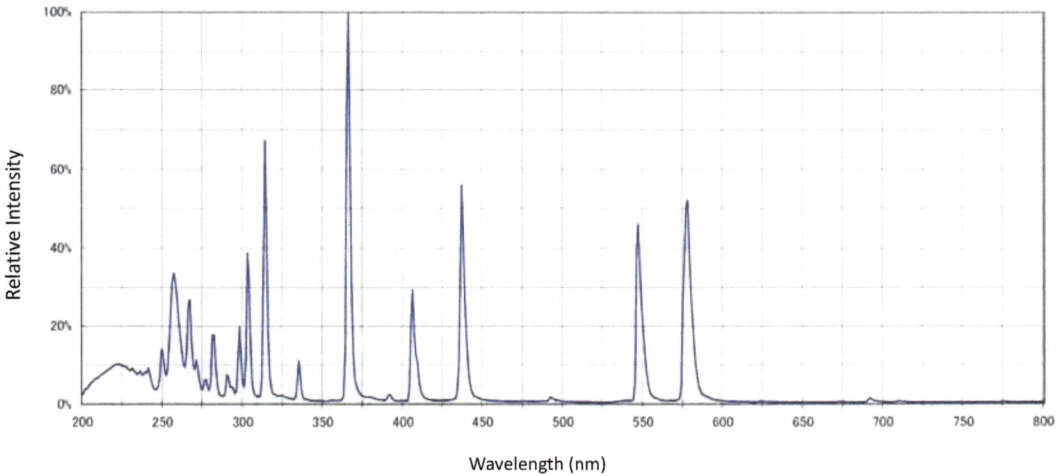

Fig. (21). Wavelength output of the medium-pressure mercury H lamp.

The wavelength output can be modified by combining and adding traces of metals to the mercury, normally called a metal halide lamp. The most commonly used trace metals are iron and gallium. Fig. (**22**) shows the wavelength output of a typical metal halide, iron-doped D lamp. As can be seen, a metal halide lamp generates stronger UVA /VV-vis output at 366 nm and 404 nm than an MPM lamp. Metal halide lamps are more effective for thick, pigmented uses.

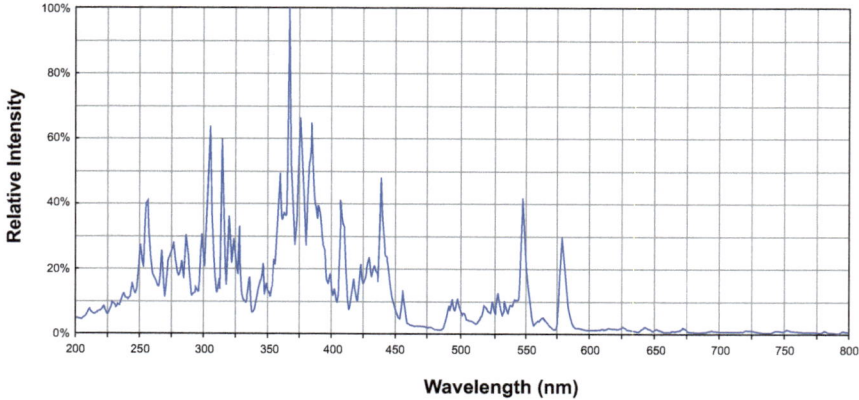

Fig. (22). Wavelength output of the metal halide lamp.

Light-emitting diode (LED) lamps can also deliver UV light. UV-LED lamps have been increasingly used in recent years. Typical commercial UV-LED lamps for UV curing use generate UV output lines at 365nm, 385 nm, 395 nm, and 405 nm. UV-LED lamp is a single wavelength output. Fig. (23) compares their wavelength output with the standard mercury lamp. The surface cure is always a big challenge for UV-LED curing formulation development as there is no short UV C light generated.

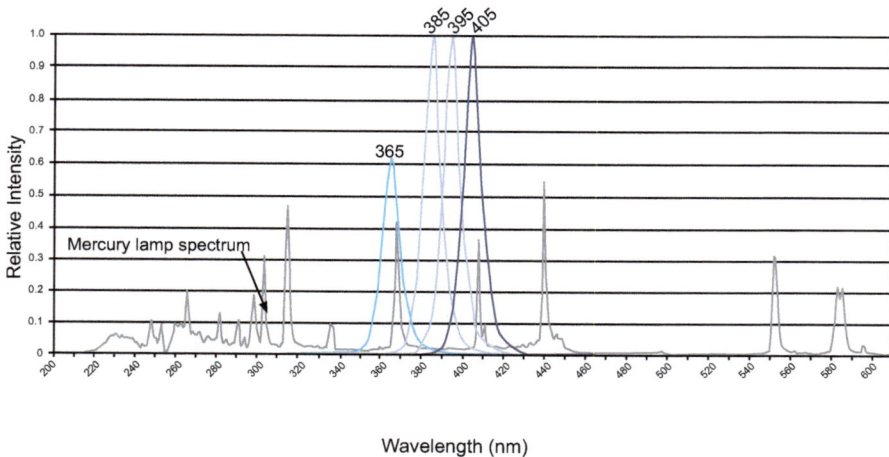

Fig. (23). Wavelength output of typical UV-LED lamps as compared to HMP lamps.

UV Cure Systems

UV cure system is typically composed of a UV lamp, reflector, cooling system, power system, and housing chamber. There are mainly three types of UV cure systems: desktop batch type, convey type, and spot type, as shown in Fig. (**24-26**), respectively. In some cases, the medium-pressure mercury lamp and metal halide lamp can be assembled into one UV cure system so the users can select suitable lamp sources for specific application requirements. Desktop batch type is widely used in UV curing applications. UV intensity can be adjusted by changing the height distance of the UV lamp. Convey-type UV cure system is especially suitable for continuous production applications. Spot type provides very high intensity over a very narrow small area, suitable for quick small-area curing.

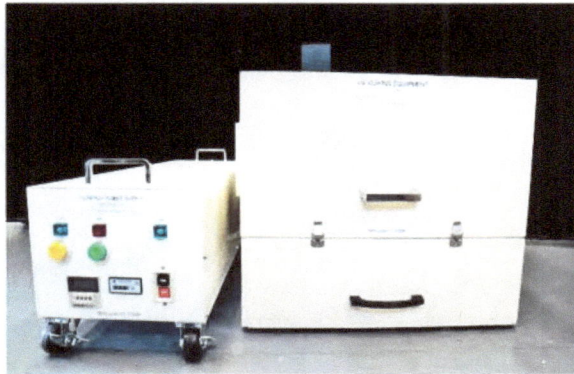

Fig. (24). Desktop flood-type UV cure system.

Fig. (25). Convey-type UV cure system.

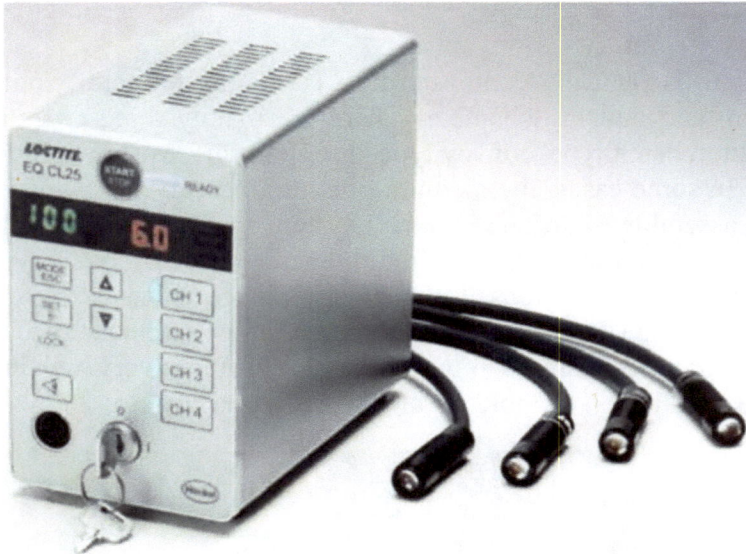

Fig. (26). Spotlight UV cure system.

UV Radiometer

UV radiometer is a device that measures the light intensity and energy associated with a specific wavelength of light. The measurement of light intensity and energy is very important in UV cure technology because they are the key factors in UV curability determination. Fig. (**27**) shows a typical commercial UV radiometer.

Light intensity at the surface is a measure of momentary exposure, often quantified in milliwatts $(mW)/cm^2$. Light energy at the surface is a measure of cumulative intensity during the exposure period of time, also called dose, quantified as millijoules $(mJ)/cm^2$.

Light intensity: A momentary exposure, measured in milliwatts per square centimeter (mW/cm^2).

UV energy (dose) (mJ/cm^2) = Intensity (mW/cm^2) x Exposure time (Seconds).

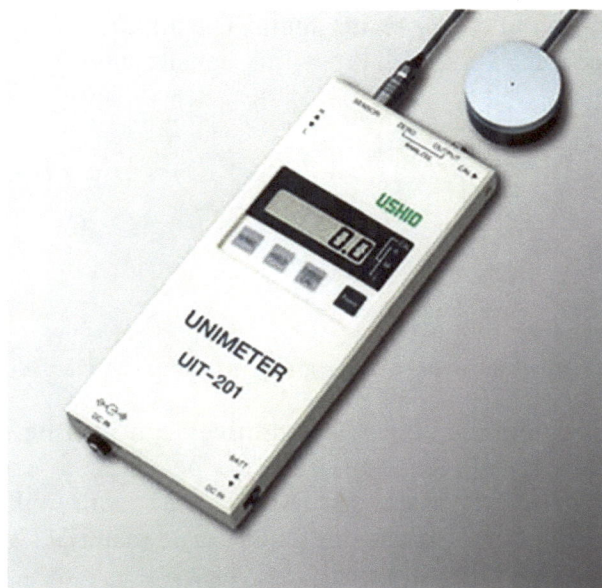

Fig. (27). A typical UV radiometer.

UV CURE CATIONIC EPOXY ADHESIVES

Formulating UV Cure Cationic Epoxy Adhesives

Epoxy resins are the adhesive base composition. Cycloaliphatic type epoxy resins are most widely used due to fast cure speed *via* cationic polymerization. Common glycidyl ether-type epoxy resins such as bisphenol A glycidyl ether and novolac glycidyl ether can also be used as these epoxy resins usually show better adhesion, although their cure speed *via* cationic polymerization is slower. Recently, hydrogenated bisphenol A epoxy resin has been used increasingly because of its faster cure speed *via* cationic polymerization and better optical properties as compared to glycidyl ether-type epoxy resins. Oxetanes can also be combined with epoxy resins for cure speed enhancement and handling property improvement.

A cationic photoinitiator is typically added with a small content in the range of 0.5 to 5%. A photosensitizer might be needed, especially for long-wavelength light sources, such as UV-LED lamp applications.

Polyols are often added as reactive additives in cationic epoxy compositions. As shown in Fig. (**28**) [21], alcohols will react with the growing carbonium cation to terminate the growing polymer and regenerate the acid, which can initiate the cationic polymerization of epoxy resins again. The addition of a suitable amount of polyols can enhance the cure speed, cross-link density, and flexibility.

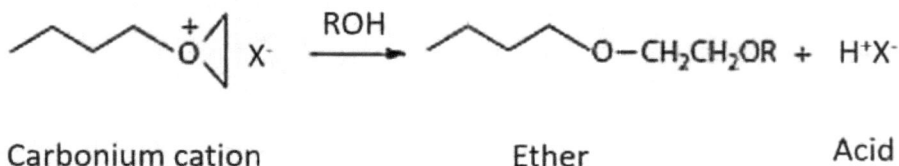

Carbonium cation   Ether   Acid

Fig. (28). Chain transfer mechanism of hydroxy group in cationic polymerization.

Fillers, tougheners, colorants, and other additives can also be used for UV-cure cationic epoxy adhesives. Precautions need to be taken to consider their light transmittance and to avoid basic additives. Cure depth will be significantly damaged in the case of poor light transmittance. Basic materials will inhibit cationic cure as they can react with the acid easily.

The function, component, main role, and typical content of UV cure cationic epoxy composition are summarized in Table **3**.

Table 3. UV cure cationic epoxy composition.

Function	Component	Main Role	Typical Content (%)
Primary	Epoxy resin	Adhesive base	20 – 95
	Cationic photoinitiator	Curability	0.5 – 5
	Photosensitizer	Curability at long wavelength	0 – 2
	Polyol	Cure speed enhancement	0 – 10
	Oxetane	Cure speed enhancement	0 – 20
Modifier	Filler	Property enhancement	0 – 70
	Toughener	Toughness enhancement	0 – 20

(Table 3) cont.....

Additive	Colorant	Coloring	0 – 2
	Coupling agent	Adhesion promotion	0 – 2
	Thixotropic agent	Rheology control	0 – 5

Properties, Testing and Characterization

The five factors below should be taken into consideration while testing UV cure cationic epoxy adhesives:

- ✓ Lamp type.
- ✓ Light intensity.
- ✓ Substrate.
- ✓ Temperature.
- ✓ Moisture.

The spectral output of the lamp type needs to be matched well with the absorbency property of the adhesive product's photoinitiators. Light intensity on the product is mainly a function of light source power, distance from the light source to the product, and the transmission property of the substrate when UV light is transmitted through. Light intensity and energy are measured with a UV radiometer for the cure profile determination and cure process control.

UV cure cationic systems are sensitive to basic materials. Basic materials will inhibit cationic polymerization, and thus, basic substrates should be avoided.

Cationic polymerization is generally a thermal-driven reaction. As illustrated in Fig. (**29**), the cure speed of cationic polymerization becomes much faster at higher temperature conditions [22]. Temperature factor needs to be considered especially when conventional lamps are used.

Water can act as a chain transfer agent like alcohols, affecting the cure speed and cure properties in UV cationic curing. As illustrated in Fig. (**30**), moisture humidity has a big impact on cure speed, especially at high humidity conditions [23].

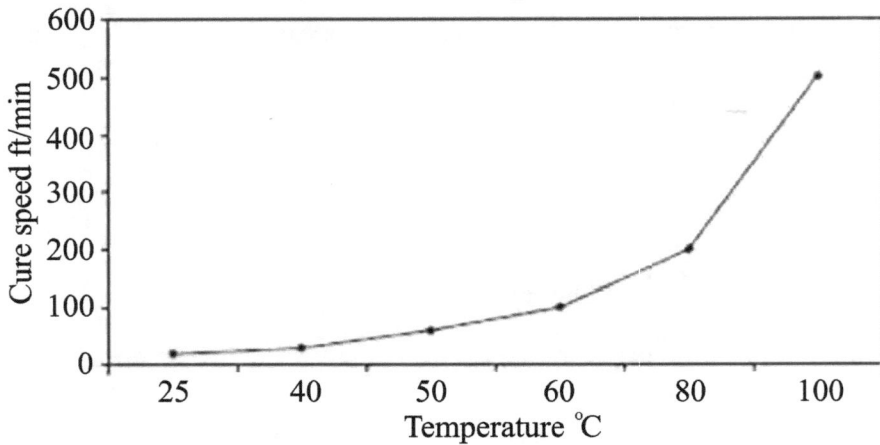

Fig. (29). Cure speed vs temperature for cationic polymerization.

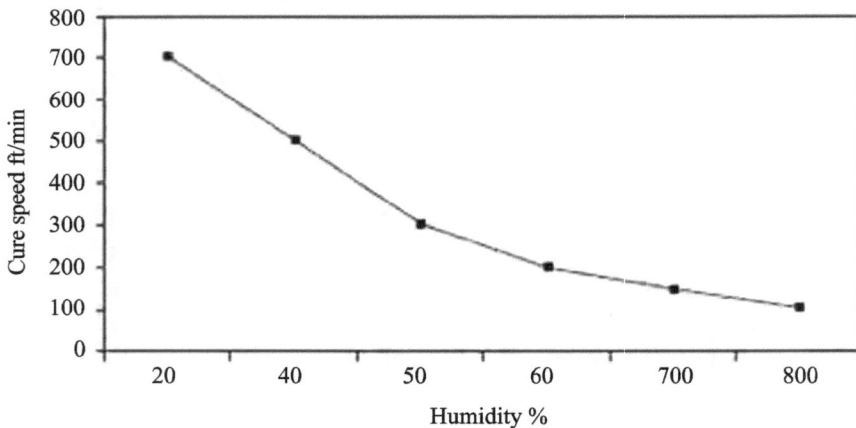

Fig. (30). Cure speed *vs* humidity for cationic polymerization.

Typical properties, testing methods, and characterization of UV cure cationic epoxy adhesives in uncured state, curing stage, and cured and adhered conditions are summarized in Table **4** [24-27]. Appearance, viscosity, and thixotropic value are the most important properties of uncured epoxy adhesives, determining primarily how to apply them on the adherend substrates. Appearance is usually measured using the visual method. Viscosity and thixotropic values can be easily measured by a viscometer.

Table 4. Properties, testing, and characterization of UV cure cationic epoxy adhesives.

Stage	Property	Test Method	Test Equipment
Uncured state	Appearance	ASTM E 284-17	Visual inspection
	Viscosity	ISO 3219, ASTM D2196	Viscometer
Curing stage	Cure properties	-	
	Cure behavior	ISO 11357-1, ASTM E 472	Photo-DSC
	Cure conversion	ISO ASTM E 1252	FT-IR
	Cure shrinkage	ISO 1675, ASTM D 792	Density meter
Cured state	Hardness	ISO 868, ASTM 2240	Hardness gauge
	Modulus	ISO 527-3, ASTM D 882	Tension tester
	CTE	ISO 11359-2, ASTM E 1545	TMA
	Tg	ISO 11359-2, ASTM E 1545	DSC, TMA, DMA
Adhesion performance	Shear strength	ISO 4587, ASTM 1002-94	Tension tester
	Tensile strength	ISO 6922, ASTM D2095-92	Tension tester
	Peel strength	ISO 11339, ASTM D1876-93	Tension tester
	Impact strength	ISO 9653, ASTM 950-94	Impact tester

Curing properties are important for handling and cure condition determination. The cure speed of UV cure cationic epoxies depends on both the product composition and actual light intensity. Any factor that reduces light intensity of the product will reduce cure speed. Depth of cure is enhanced by a longer wavelength as it can penetrate deeper into the product. Depth of cure is a function of product composition, wavelength, light intensity, and exposure time. Surface cure, in contrast, is enhanced by shorter, more energetic wavelengths. The cure behavior of light cure materials can be quantitatively analyzed by differential scanning photo calorimetry (Photo-DSC) to measure changes in enthalpy generated by cationic polymerization during or after UV light exposure. FT-IR is often used to analyze cure conversion or degree of cure of epoxy/oxetane composition by measuring changes in characteristic absorption peak area of the epoxy group and oxetane group during and after light exposure. Characteristic absorption of epoxy and oxetane group is summarized in Table **5** [28]. Low curing shrinkage is one key feature of UV cure cationic epoxy adhesives. It is normally calculated from the density measurement of uncured and cured epoxy adhesives.

Table 5. FT-IR characteristic absorption peak of epoxy and oxetane groups.

Functional Group	Characteristic Absorption
Epoxy group, glycidyl ether type	916 cm^{-1}
Epoxy group, cycloaliphatic type	789 cm^{-1}
Oxetane group, oxetanes	995 cm^{-1}

The physical, thermal, and mechanical properties of cured epoxy adhesives can be measured by preparing suitable cured samples with the use of specific testing and analytical methods. Hardness is measured by a hardness tester. CTE (coefficient of thermal expansion) is measured by the TMA (thermal mechanical analysis) method. Tg (glass transition temperature) is the key factor, representing temperature-resistant range of epoxy adhesives. Tg can be measured by various thermal analysis methods such as DSC, TMA, and DMA (dynamic mechanical analysis). Mechanical properties such as tensile modulus and strength can be measured by a tension tester. Its storage modulus is often measured by DMA. Other properties of cured epoxy adhesives, such as electrical properties and optical properties, can also be measured with the use of specific testing methods upon necessary.

There are three main tests to determine the adhesion strength of epoxy adhesives – tensile, shear, and peel. Tensile strength is the resistance of a material to breaking under tension testing, representing the adhesion behavior of an adhesive material while an axial stretching load is applied. Shear strength, commonly called lap shear strength, is the load that a material can withstand in a direction parallel to the face of the material, representing the maximum shear stress in the adhesive prior to failure under torsional loading. Peel strength is the average force required to decompose two adhered substrates. Peel strength is the average load per unit width of bond line, while shear and tensile strength is the peak load per adhesion area.

Typical Applications

UV cure cationic epoxy adhesives can be cured quickly and have been very successfully used in electronics assembly and general bonding applications such as general glass bonding, optical parts assembly, image sensor module assembly, display panel bonding, and module assembly that require fast production speed and high adhesion performance [29-36].

General Purpose Bonding

UV cure cationic epoxy adhesive is primarily used for transparent material bonding applications such as general glass bonding, transparent plastic bonding, and optical module assembly. Compared to normal UV acrylate adhesive, UV cure cationic epoxy adhesive shows better adhesion performance and can withstand strict environmental testing conditions. By combining the use of a multifunctional epoxy silane coupling agent with bisphenol A epoxy resin, a high-performance UV cure cationic epoxy composition is shown in Table **6** [37]. As can be seen, sample no. 1 composition shows outstanding adhesion performance at very strict autoclave humidity conditions.

Table 6. High-performance UV cure epoxy adhesive.

Sample No.	1	2
Composition, %		
D.E.R. 331[1]	92	-
D.E.R. 383[2]	-	92.4
Multifunctional epoxy silane	5	4.6
3-Glydoxypropyltrimethoxysilane	-	-
CPI-6976[3]	3	3
Appearance	Clear liquid	Clear liquid
Viscosity, mPa.s/25°C	12000	4200
Glass adhesion strength[4], MPa		

(Table 6) cont.....

Initial	1.30	1.40
96 hours @105°C100%	1.40	0
12 hours @121°C 2atm	0.50	0

1. Bisphenol A epoxy resin (EEW=190), supplied by Olin Corporation.

2. Bisphenol A epoxy resin (EEW=180), supplied by Olin Corporation.

3. Cationic photoinitiator, supplied by Aceto Corporation.

4. Measured according to ASTM 6922. Substrate: optical glass slide.

Sample cure condition: 60 seconds @100 mW/cm^2, medium-pressure mercury H lamp used.

Image Sensor Package Sealing

UV cure cationic epoxy adhesives have been successfully used to bond and seal cover glass and the sensor package in digital camera assembly applications because of fast curability and high adhesion performance. We have found that the adhesion performance of UV cationic epoxy adhesives can be much improved by using a combination of cationic photoinitiator with thermal cationic initiator [38, 39]. As summarized in Table **7**, samples with cationic thermo-initiator and talc filler performed well on the crosshatch tape testing at both initial and boiling water immersion conditions. The adhesive product developed based on this study showed excellent reliability performance in CMOS (complementary metal-oxide semiconductor) image sensor package sealing application.

Table 7. UV cationic epoxy composition study for image sensor packaging.

Sample No.	1	2	3	4	5
Component, %					
Bisphenol A epoxy resin[1]	-	34.5	69.5	69.5	86.0
Hydrogenated bisphenol A epoxy resin[2]	65.5	34.0	-	-	-
	1.1	1.1	1.1	1.6	1.0

(Table 7) cont.....

Cationic photo-initiator[3]	0.5	0.5	0.5	-	-
Cationic thermo-initiator[4]	5.0	5.0	6.0	6.0	10.0
Polyol	25.0	22.0	19.0	19.0	-
Talc	2.9	2.9	2.9	2.9	1.0
Silane coupling agent	-	-	1.0	1.0	1.0
Fumed silica					
Minimum UV energy needed, mJ/cm^2	1200	1800	3600	3600	3600
Tg[5], °C	86	107	123	-	-
Crosshatch tape test[6]					
Initial	100/100	100/100	100/100	100/100	0/100
Boiling water immersion x 4 hours	100/100	100/100	100/100	30/100	-

*1. RE-310S, manufactured by Nippon Kayaku Co., Ltd.

2. Epiclon EXA-7015, manufactured by DIC Corp.

3. Cyracure UVI-6990, manufactured by Union Carbide Inc.

4. San-Aid SI-60L, manufactured by Shanshin Chemical Industrial Co., Ltd.

5. Measured by TMA method. Sample cure condition: UV 3600 mJ/cm^2 + 120°C for 60 min.

6. Measured according to ASTM 1002-94. Substrate: glass slide.

Sample cure condition: UV 3600 mJ/cm^2 + 120°C for 60 min.

Delay Cure Technology

UV cure adhesive is typically limited to bond transparent materials. This is because it cannot be cured due to the lack of UV light needed for active species formation. Uncure problems always occur in shadow areas where UV light can not be passed through in the use of UV cure materials. UV delay cure technology has been developed to resolve this issue. As illustrated in Fig. (**31**) [40], the delay cure type adhesive will remain at an uncured state after UV light exposure for a certain period of time (called as open time), so assembly of the other substrate can be applied after the UV light exposure. It will start to cure after the open time and can also be cured quickly under heating conditions.

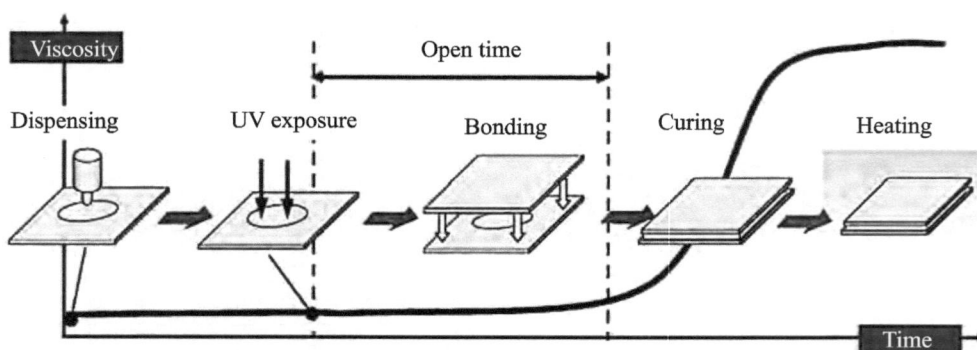

Fig. (31). Application and cure process of delay cure adhesive.

Delay cure technology is based on UV cure cationic epoxy composition with a combination use of a suitable amount of retardant chemicals such as crown ether, a weak basic chemical that can capture and release the Bronsted acid for a certain period of time [41-44]. Delay cure epoxy compositions and their cure behavior are illustrated in Tables **8**, **9** shows the product properties and performance of a commercial delay cure epoxy product supplied by Sekisui Chemical Co., Ltd [45].

Table 8. UV cure epoxy composition.

Sample	Delay cure I	Delay cure II	Normal
Composition, %			
Epiclon EXA-830LVP[1]	50	-	50
jER 8034[2]	25	-	25
jER YL6753[3]	25	-	25
Celloxide 2021[4]	-	100	1
DTS-200[5]	1	1	0.01

(Table 8) cont.....

DETX-S[6]	0.01	0.01	-
jER 630[7]	0.25	0.25	1
KBM-403[8]	1	1	1
Surflon S-611[9]	1	1	
Viscosity, mPa.s/25°C	500	400	500
Delay cure result[10] 10 minutes after UV radiation Post thermal cure	 Uncured Cured	 Uncured Cured	 Cured -

1. Bisphenol F epoxy resin, supplied by DIC Corporation.

2. Hydrogenated bisphenol A epoxy resin, supplied by Mitsubishi Chemical Corporation.

3. Hydrogenated bisphenol F epoxy resin, supplied by Mitsubishi Chemical Corporation.

4. Cycloaliphatic epoxy resin (ECC), supplied by Daicel Corporation.

5. Sulphonium type cationic photoinitiator, supplied by Midori Kagaku Co., Ltd.

6. Photosensitizer, supplied by Nippon Kayaku Co., Ltd.

7. Triglycidyl p-aminophenol, supplied by Mitsubishi Chemical Corporation.

8. Silane coupling agent, supplied by Shin-Estu Chemical Co., Ltd.

9. Leveling agent, supplied by AGC Semi Chemical Co., Ltd.

Table 9. Typical properties of delay cure epoxy adhesive.

Appearance	Colorless liquid
Viscosity, mPa.s/25°C	470
UV cure energy, mJ/cm^2	1500
Open time, min @25°C	15

(Table 9) cont.....

Post thermal cure	80°C x 30 min
Adhesion strength, MPa/glass	4.1
Tg, by DMA	92°C

CONCLUSION

Cycloaliphatic type epoxy resins are most widely used in UV cationic epoxy formulations due to fast cure speed *via* cationic polymerization. Oxetanes can also be combined together with epoxy resins to cure speed enhancement and handle property adjustment. Sulfonium salts with SbF_6^- and PF_6^- are the most used UV cationic photoinitiators. Iodonium salts are often used with a combination of photosensitizers. Medium-pressure mercury lamps and metal halide lamps are widely used UV lamps. UV-LED lamps have been used increasingly in recent years. Lamp type, light intensity, substrate, temperature, and moisture should be taken into consideration when testing UV cure cationic epoxy adhesives. Basic materials will inhibit cationic cure due to easy reaction with acid, the cationic species. UV cure cationic epoxy adhesives can be cured quickly *via* UV radiation and have been very successful in electronics assembly and general bonding applications such as general glass bonding, optical parts assembly, image sensor module assembly, display panel bonding, and module assembly that require fast production speed and high adhesion performance.

REFERENCES

[1]   Voytekunas, V.Y.; Ng, F.L.; Abadie, M.J.M. Kinetics study of the UV-initiated cationic polymerization of cycloaliphatic diepoxide resins. *Eur. Polym. J.,* **2008**, *44*(11), 3640-3649.

http://dx.doi.org/10.1016/j.eurpolymj.2008.08.043

[2]   Golaz, B.; Michaud, V.; Leterrier, Y.; Månson, J.A.E. UV intensity, temperature and dark-curing effects in cationic photo-polymerization of a cycloaliphatic epoxy resin. *Polymer (Guildf.),* **2012**, *53*(10), 2038-2048.

http://dx.doi.org/10.1016/j.polymer.2012.03.025

[3]   Chen, C. Structural epoxy adhesives, in: *Structural Adhesives*: Properties, Characterization and Applications; Mittal, K.L. and Panigraphi, S.K., Eds.; Wiley – Scrivener: Beverly, MA, **2023**, pp 3-30.

http://dx.doi.org/ 10.1002/9781394175604.ch1

[4]   Chen, C.; Li, B.; Wang, C.; Iwasaki, S.; Kanari, M.; Lu, D. UV and thermal cure epoxy adhesive. *Painting and Coatings Industry*; Yalmaz, F., Ed.; IntechOpen: London, UK, **2019**, pp. 71-85.
        http://dx.doi.org/10.5772/intechopen.82168

[5]   Fazal-ur-Rehman, M. Photo-curing schemes to cure the epoxy resins and their impacts on curing process. *Journal of chemistry and applications,* **2018**, *4*(1), 01-05.
        http://dx.doi.org/10.13188/2380-5021.1000010

[6]   Dekker, C. UV-radiation curing of adhesives.*Adhesives and Sealants*; Cognard, P., Ed.; Elsevier: London, **2006**, pp. 309-340.

[7]   Sangermano, M.; Razza, N.; Crivello, J.V. Cationic UV curing: technology and applications. *Macromol. Mater. Eng.,* **2014**, *299*(7), 775-793.
        http://dx.doi.org/10.1002/mame.201300349

[8]   Sasaki, H. Curing properties of cycloaliphatic epoxy derivatives. *Prog. Org. Coat.,* **2007**, *58*(2-3), 227-230.
        http://dx.doi.org/10.1016/j.porgcoat.2006.09.030

[9]   Noè, C.; Hakkarainen, M.; Sangermano, M. Cationic UV-curing of epoxidized biobased resins. *Polymers (Basel),* **2020**, *13*(1), 89.
        http://dx.doi.org/10.3390/polym13010089 PMID: 33379390

[10]   Xia, Y.; Zhang, D.; Li, Z.; Lin, H.; Chen, X.; Oliver, S.; Shi, S.; Lei, L. Toughness modification of cationic UV-cured cycloaliphatic epoxy resin by hydroxyl polymers with different structures. *Eur. Polym. J.,* **2020**, *127*, 109594.
        http://dx.doi.org/10.1016/j.eurpolymj.2020.109594

[11]   Lu, W.Y.; Bhattacharjee, S.; Lai, B.X.; Duh, A.B.; Wang, P.C.; Tan, C.S. Hydrogenation of bisphenol A-type epoxy resin (BE186) over vulcan XC72-supported Rh and Rh–Pt catalysts in ethyl acetate-containing water. *Ind. Eng. Chem. Res.,* **2019**, *58*(36), 16326-16337.
        http://dx.doi.org/10.1021/acs.iecr.9b02583

[12]   Esposito Corcione, C.; Malucelli, G.; Frigione, M.; Maffezzoli, A. UV-curable epoxy systems containing hyperbranched polymers: Kinetics investigation by photo-DSC and real-time FT-IR experiments. *Polym. Test.,* **2009**, *28*(2), 157-164.
        http://dx.doi.org/10.1016/j.polymertesting.2008.11.002

[13]   Green, W.A. *Industrial phtoinitiator*; CRC Press: New York, **2010**, pp. 150-151.
        http://dx.doi.org/10.1201/9781439827468

[14]   Speedcure 976, Arkema Corporation. https://emea.sartomer.arkema.com/en/product-finders/product/f/sartomer_Lambson/p/speedcure-976/ [Accessed on December 20, 2023].

[15]   Green, W.A. *Industrial phtoinitiator*; CRC Press: New York, **2010**, p. 152.
        http://dx.doi.org/10.1201/9781439827468

[16]   Speedcure 937, Arkema Corporation. https://emea.sartomer.arkema.com/en/product-finders/product/f/sartomer_Lambson/p/speedcure-937/ [Accessed on December 20, 2023].

[17]   Speedcure938, Arkema Corporation. https://emea.sartomer.arkema.com/en/product-finders/product/f/sartomer_Lambson/p/speedcure-938/ [Accessed on December 20, 2023].

[18]   Green, W.A. *Industrial phtoinitiator*; CRC Press: New York, **2010**, p. 169.
http://dx.doi.org/10.1201/9781439827468

[19]   *Light cure adhesive technology guide;* Henkel Corporation. https://dm.henkel-dam.com/is/content/henkel/lt-2730-brochure-light-cure-adhesive-technology-guide [Accessed on December 20, 2023].

[20]   *UV cure system*; Jingke Printing System Co., Ltd. https://www.jkuv.org/uv-lamp-work.html [Accessed on December 20, 2023].

[21]   Green, W.A. *Industrial phtoinitiator*; CRC Press: New York, **2010**, p. 154.
http://dx.doi.org/10.1201/9781439827468

[22]   Popp, M.; Harwig, A.; Teczyk, K. Influence of temperature and atmosphere on the curing rate and final properties of UV curing acrylate and epoxides. *RadTech Eu. Conf. Proc,* **1989**, pp. 607-614.

[23]   Brann, W.L. The effects of moisture on UV curable cationic epoxide sustems. *RadTech Eu. Conf. Proc,* **1989**, pp. 565-579.

[24]   *Plastics—Epoxy—Test Method*, ISO 18280: 2010; International Organization for Standardization: **2010**.

[25]   *Standard Guide for Testing Epoxy Resins*, ASTM D 4142-89; ASTM International: West Conshohocken, PA, **2009**.

[26]   Standard specification for epoxy adhesive for bonding metallic and nonmetallic materials. *ASTM D,* **2012**, *6412*(D6412), M-99.

[27]   Biron, M. *Thermosets and Composites,* 2nd ed; Elsevier: London, **2014**, pp. 211-212.

[28]   Tanaka, H. Cure behavior analysis for UV cure resins.*Optimization on UV cure process*; Fukushima, H., Ed.; Science and Technology Publishing: Tokyo, **2008**, pp. 50-64.

[29]   Chiang, T.H.; Hsieh, T.E. A study of monomer's effect on adhesion strength of UV-curable resins. *Int. J. Adhes. Adhes.,* **2006**, *26*(7), 520-531.
http://dx.doi.org/10.1016/j.ijadhadh.2005.07.004

[30]   Goss, B. Bonding glass and other substrates with UV curing adhesives. *Int. J. Adhes. Adhes.,* **2002**, *22*(5), 405-408.
http://dx.doi.org/10.1016/S0143-7496(02)00022-2

[31]   Velankar, S.; Pazos, J.; Cooper, S.L. High-performance UV-curable urethane acrylates *via* deblocking chemistry. *J. Appl. Polym. Sci.,* **1996**, *62*(9), 1361-1376.
http://dx.doi.org/10.1002/(SICI)1097-4628(19961128)62:9<1361::AID-APP6>3.0.CO;2-F

[32]   Fourassier, J.; Lalevee, J. *Photoinitiator for Polymer Synthesis*; Wiley-VCH Verlag: Weinheim, Germany, **2012**, pp. 41-72.
http://dx.doi.org/10.1002/9783527648245.ch4

[33]   Iida, T. Troubleshooting of UV adhesives for opto-electronics. *J. Journal of The Adhesion Society of Japan,* **2007**, *43*(2), 72-78.

http://dx.doi.org/10.11618/adhesion.43.72

[34] Ramasamy, E.; Karthikeyan, V.; Rameshkumar, K.; Veerappan, G. Glass-to-glass encapsulation with ultraviolet light curable epoxy edge sealing for stable perovskite solar cells. *Mater. Lett.,* **2019**, *250*, 51-54.

http://dx.doi.org/10.1016/j.matlet.2019.04.082

[35] Lu, Q.; Yang, Z.; Meng, X.; Yue, Y.; Ahmad, M.A.; Zhang, W.; Zhang, S.; Zhang, Y.; Liu, Z.; Chen, W. A Review on encapsulant technology from organic light emitting diodes to organic and perovskite solar cells. *Adv. Funct. Mater.,* **2021**, *31*(23), 2100151.

http://dx.doi.org/10.1002/adfm.202100151

[36] Doerfler, R.; Barth, S.; Boeffel, C.; Wedel, A. New UV-curing OLED encapsulation adhesive with low water permeation. *SID Int. Symp. Dig. Tec.,* **2006**, pp. 440-443.

http://dx.doi.org/10.1889/1.2433526

[37] Chen, C.; Li, B.; Wang, C.; Iwasaki, S.; Kanari, M. Light cure epoxy composition. *Chinese Patent 110582540,* **2022**.

[38] Gan, Y.; Chen, C.; Terada, K. Cationically photocurable epoxy resin composition. *U.S. Patent 7456230,* **2002**.

[39] Chen, C.; Gan, Y. Cationically curable epoxy resin composition. *U.S. Patent 7795744,* **2010**.

[40] Shichiri, T. UV delay cure epoxy resin. *Journal of Network Polymer, Japan,* **2021**, *42*(3), 105-112.

[41] Watanabe, Y.; Shichiri, T. Post-curable light resin composition. *Japanese Patent 5799177,* **2015**.

[42] Suzuki, K.; Takazawa, Y. Light cure epoxy composition. *Japanese Patent 5364460,* **2013**.

[43] Gunther, E.; Gerald, U. Cationically hardening multicomponent epoxy resin compositions and a process for their cure. *Eur. Patent Appl. 0688804,* **1995**.

[44] Teraguchi, Y.; Shichiri, T. Optical post-curable resin composition. *Japanese Patent Appl.* 2015-44917, **2015**.

[45] Sekisui Chemical Co., Ltd. *UV Delayed Curing Low Moisture Permeable Adhesive Photolec E*; Sekisui Chemical Co., Ltd.: https://www.sekisui.co.jp/electronics/en/resin /photolec-e.html (accessed Dec 20, 2023)].

CHAPTER 3

Dual Cure Hybrid Epoxy Technology

Abstract: UV and thermal dual cure hybrid epoxy adhesives are composed of partially acrylated epoxy resin, acrylates, free radical photoinitiators, epoxy resins, and latent curing agents with modifiers and additives. The acrylate components are cured momently *via* light radiation and thus can instantly fix the bonding materials at room temperature. Their epoxy components can be cured at the post-thermal cure stage to achieve satisfactory adhesion performance. Dual cure hybrid epoxy technology has been very successfully used in electronics assembly and general bonding applications.

Keywords: Acrylate monomer, Acrylate oligomer, Cure shrinkage, Dual cure, Free radical photoinitiator, Oxygen inhibition, Partially acrylated epoxy resin.

UV CURE ACRYLATE CHEMISTRY

UV and thermal dual cure hybrid epoxy adhesives combine UV cure acrylate composition with thermal cure epoxy composition. Dual cure hybrid epoxy adhesives are primarily composed of partially acrylated epoxy resin, acrylates, free radical photoinitiators, epoxy resin, and latent curing agents. Their acrylate components can be cured instantly *via* light radiation, and thus, the bonding materials can be fixed momently at room temperature. Their epoxy components are cured at the post-thermal cure stage to achieve satisfactory adhesion reliability performance. Dual cure hybrid epoxy technology has been very successfully used in general structural bonding, camera module assembly, and LCD panel bonding applications. The dual cure hybrid epoxy adhesives combine the advantages of instant curability of UV acrylate composition with a high-reliability performance of thermal cure epoxy part very well.

Most widely used UV cure adhesives are acrylate-based [1-8]. Acrylate-based UV cure adhesives are primarily composed of acrylate monomer, acrylate oligomer, and photoinitiator. The photoinitiator will generate free radicals *via* UV radiation, initiating free radical polymerization of acrylate compositions rapidly. Acrylate-based UV cure adhesives can be cured instantly within a few seconds *via* UV light radiation. Surface cure issue, shadow cure problem, relatively high cure shrinkage, and poor humidity reliability are their main limitations.

Acrylate Monomers

Acrylate monomer is the fundamental component of UV cure acrylate formulation technology. Acrylate monomer plays a key role in the determination of its cure behavior, handling property, and physical property and affects its adhesion performance. Acrylate monomers can be classified as mono-functional, di-functional, tri-functional, and multi-functional acrylates.

Chemical structure, viscosity, CAS No., molecular weight, and Tg of typical mono-functional acrylate and methacrylate monomers are shown in Fig. (**1**). Tg is their homopolymer's value.

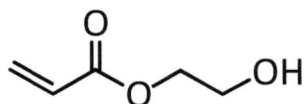

Viscosity: 6 mPa.s/20°C
CAS. No.: 818-61-1
Molar mass: 116
Tg: -15°C

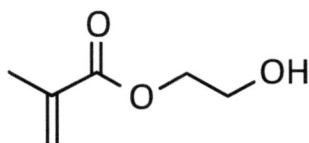

2-Hydroxyethyl acrylate

Viscosity: 6.8 mPa.s/20°C
CAS. No.: 868-77-9
Molar mass: 130
Tg: 25°C

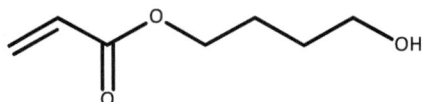

2-hydroxyethyl methacrylate

Viscosity: 5.5 mPa.s/25°C
CAS. No.: 2478-10-6
Molar mass: 144
Tg: -32°C

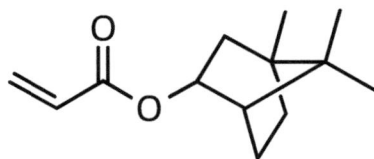

4-Hydroxybutyl acrylate

Viscosity: 7.7 mPa.s/20°C
CAS. No.: 5888-33-5
Molar mass: 208
Tg: 97°C

Isobornyl acrylate

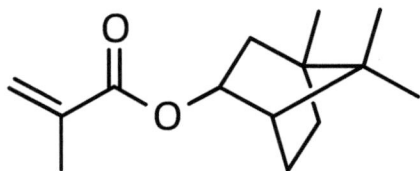

Viscosity: 6 mPa.s/25°C
CAS. No.: 7534-94-3
Molar mass: 222
Tg: 180°C

Isobornyl methacrylate

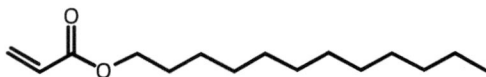

Viscosity: 4 mPa.s/25°C
CAS. No.: 2156-97-0
Molar mass: 240
Tg: 15°C

Lauryl acrylate

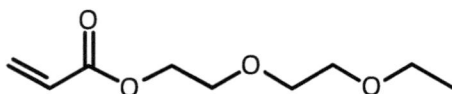

Viscosity: 2.9 mPa.s/25°C
CAS. No.: 7328-17-8
Molar mass: 188
Tg: -67°C

Ethoxyethoxyethyl acrylate

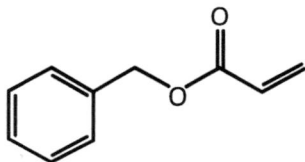

Viscosity: 2 mPa.s/25°C
CAS. No.: 2495-35-4
EEW: 162
Tg: 6°C

Benzyl acrylate

Fig. (1). Typical monofunctional acrylate monomers and methacrylate monomers.

Chemical structure, viscosity, CAS No., molecular weight, and Tg of typical di-, tri-, and multi-functional acrylate and methacrylate monomers are shown in Fig. (**2**). The combination use of these acrylates can increase the degree of cross-linking of the polymer matrix.

Viscosity: 135 mPa.s/25°C
CAS. No.: 42594-17-2
EEW: 304
Tg: 186°C

Dimethylol triclodecane diacrylate

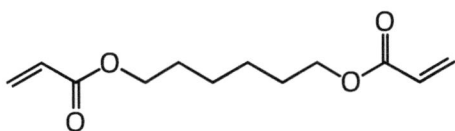

1,6-dihexanediol diacrylate

Viscosity: 7 mPa.s/25°C
CAS. No.: 13048-33-4
EEW: 226
Tg: 43°C

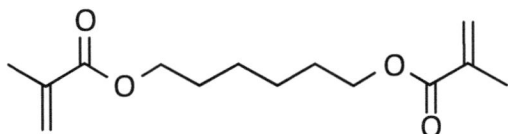

1,6-Hexanediol dimethacrylate

Viscosity: 8 mPa.s/25°C
CAS. No.: 6606-59-3
EEW: 254
Tg: 30°C

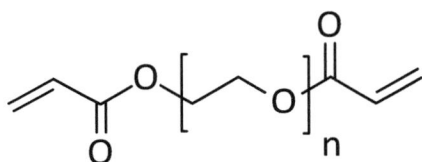

Polyethylene glycol (400) diacrylate

Viscosity: 57 mPa.s/25°C
CAS. No.: 26570-48-9
EEW: 508
Tg: -25°C

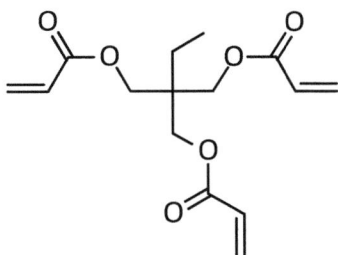

Trimethylolpropane triacryalte

Viscosity: 106 mPa.s/25°C
CAS. No.: 15625-89-5
EEW: 296
Tg: 62°C

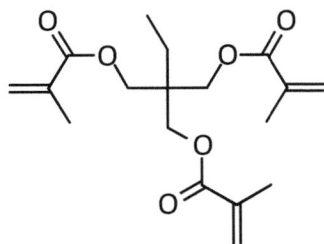

Trimethylolpropane trimethacrylate

Viscosity: 44 mPa.s/25°C
CAS. No.: 3290-92-4
EEW: 338
Tg: 27°C

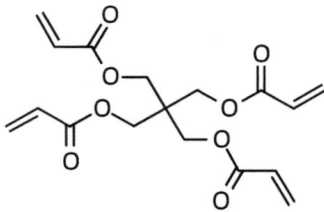

Viscosity: 342 mPa.s/25°C
CAS. No.: 4986-89-4
EEW: 352
Tg: 103°C

Pentaerythritol tetraacrylate

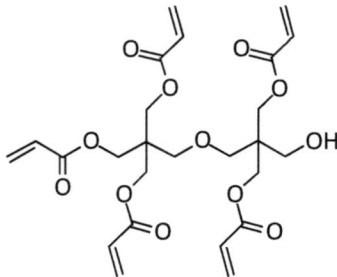

Viscosity: 13600 mPa.s/25°C
CAS. No.: 60506-81-2
EEW: 525
Tg: 90°C

Dipentaerylthritol petaacrylate

Fig. (2). Typical di-, tri-, and multi-functional acrylate monomers and methacrylate monomers.

Cure shrinkage of acrylates occurs during the curing process due to double-bond polymerization. High cure shrinkage induces considerable stresses that cause serious adhesion failures, such as cracking, delamination, and debonding. Acrylate monomer is the main component for cure shrinkage determination due to its relatively high double-bond content. Cure shrinkage relates to functionality, molecular weight, and density of acrylates. The calculation of thermotical maximum shrinkage has been studied and proposed by Painter *et al.*, as shown in Fig. **(3)** [9]. The maximum cure shrinkage of typical acrylate monomers is summarized in Table **1**. As can be seen, most acrylate monomers have a curing shrinkage of over 10%. Actual cure shrinkage might be slightly lower because the conversion of the double bond is always less than 100 percent.

Maximum shrinkage (%) = -1.38 + 2668 x Functionality x Density / Molar mass

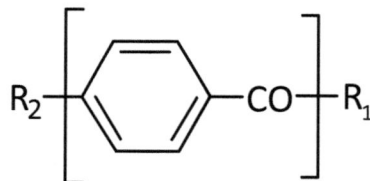

Fig. (3). Theoretical maximum shrinkage calculation of acrylates.

Table 1. Calculated maximum cure shrinkage of acrylate monomers.

Monomer	MW	Density	Functionality	Maximum Shrinkage (%)
2-Hydroxyethyl acrylate	116	1.11	1	24
2-hydroxyethyl methacrylate	130	1.07	1	20
4-Hydroxybutyl acrylate	144	1.04	1	19
Isobornyl acrylate	208	0.987	1	11
Isobornyl methacrylate	222	0.98	1	10
Lauryl acrylate	240	0.875	1	8
Ethoxyethoxyethyl acrylate	188	1.013	1	13
Benzyl acrylate	162	1.08	1	16
Dimethylol triclodecane diacrylate	304	1.1	2	18
1,6-Dihexanediol diacrylate	226	1.02	2	23
1,6-Dihexanediol dimethacrylate	254	0.982	2	19
Polyethyleneglycol diacrylate	471	1.117	2	11
Trimethylolpropane triacrylte	296	1.109	3	29
Trimethylolpropane trimethacrylte	338	1.061	3	24
Pentaerythritol tetraacrylate	352	1.179	4	34
Dipentaerylthritol petaacrylate	525	1.192	5	29

The cure speed of acrylate monomers depends a lot on their chemical structure. Methacrylate monomers cure slower than acrylate monomers due to steric hindrance. On the other hand, methacrylate polymers have higher Tg. We have studied the cure behavior of a novel photocurable material using UPLC-Q-TOF-MS and found that the UV cure speed of acrylate monomers is much different [10]. The cure speed of the studied acrylate monomers decreases in the sequence of:

4- Hydroxy- acrylate > di- and tri-functional acrylates> Isobornyl acrylate > benzyl acrylate.

Acrylate Oligomers

Urethane Acrylate Oligomers

Urethane acrylate oligomer is the most widely used acrylate oligomer. As illustrated in Fig. (4) [11], urethane acrylate oligomer is prepared in two steps: 1) Synthesis of urethane prepolymer from a diisocyanate and a polyol, typically at a 2:1 molar ratio; 2) Synthesis of urethane acrylate oligomer from the synthesized urethane prepolymer with a monofunctional hydroxy bearing acrylate monomer, typically 2-hydroxyethyl acrylate or 2-hydroxyethyl methacrylate. Various types of urethane acrylate oligomers have been commercialized and supplied to the market. Commercial-grade urethane acrylate oligomer products are sometimes diluted with a certain amount of acrylate monomers to lower viscosity for easy handling.

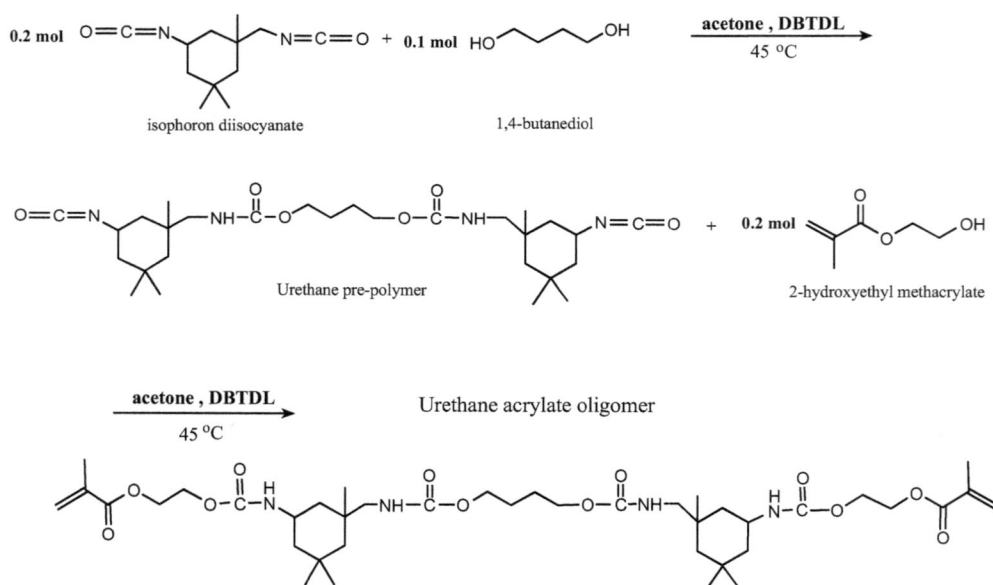

Fig. (4). Synthesis of urethane acrylate oligomer from IPDI, 1,4-butadiol and 2-HEMA.

Urethane acrylate oligomers can be classified as aromatic and aliphatic urethane acrylate oligomers depending on the aromatic or aliphatic type diisocyanate selected in the synthesis of urethane acrylate oligomers. Aromatic urethane acrylate oligomer has higher Tg but is relatively brittle and has a yellowing tendency, especially in outdoor application conditions. TDI (diisocyanate) is commonly used for the preparation of aromatic urethane acrylate monomer because of its light color, ease of use, and lower viscosity formation. Aliphatic urethane acrylate oligomer has good optical properties and is more flexible. The definition and key features of aromatic and aliphatic urethane acrylates are compared in Table **2**.

Table 2. Aromatic and aliphatic urethane acrylate oligomer.

-	Aromatic Urethane	Aliphatic Urethane
Structure	Aromatic ring contained	Nonaromatic
Isocyanate origin	TDI (toluene diisocyanate) MDI (diphenylmethane diisocyanate) NDI (naphthalene diisocyanate)	IPDI (isophorone diisocyanate) HMDI (hydrogenated MDI) HDI (hexamethylene diisocyanate)
Optical property	Yellowing tendency	Non-yellowing
Flexibility	Rigid	Flexible
Thermal resistance	Good	Moderate

Epoxy Acrylate Oligomers

Epoxy acrylate oligomer is prepared by the reaction of epoxy resin and acrylic acid, as illustrated in Fig. (**5**) [12]. Epoxy acrylate has a faster cure speed, possesses excellent chemical resistance, and shows better reliability performance than the common urethane acrylate oligomer. Epoxy acrylate oligomers are more suitable for use as dual-cure hybrid epoxy adhesives.

$$CH_2\text{-}CH\text{-}CH_2\text{-}O \overset{O}{\diagup\!\diagdown} \!\!-\!\!\bigcirc\!\!-\!\!\underset{\underset{CH_3}{|}}{\overset{\overset{CH_3}{|}}{C}}\!\!-\!\!\bigcirc\!\!-\!\!O\text{-}CH_2\text{-}CH\text{-}CH_2 \quad + 2 \quad CH_2\text{=}CH\text{-}COOH$$

$$CH_2\text{=}CH\text{-}\overset{O}{\overset{||}{C}}\text{-}O\text{-}CH_2\text{-}\overset{\overset{OH}{|}}{CH}\text{-}CH_2\text{-}O\!-\!\bigcirc\!\!-\!\!\underset{\underset{CH_3}{|}}{\overset{\overset{CH_3}{|}}{C}}\!\!-\!\!\bigcirc\!\!-\!\!O\text{-}CH_2\text{-}\overset{\overset{OH}{|}}{CH}\text{-}CH_2\text{-}O\text{-}\overset{O}{\overset{||}{C}}\text{-}CH\text{=}CH_2$$

Fig. (5). Synthesis of epoxy acrylate from bisphenol A epoxy resin and acrylic acid.

Free Radical Photoinitiators

Most commercialized free radical photoinitiators are aryl ketones, whose structure is shown in Fig. (**6**). Photoinitiators can absorb UV energy to form an excited state that will generate free radicals to initiate the polymerization of acrylate monomers and oligomers. The substitution of R_2 will influence the wavelength of light absorption, and the substitution of R_1 will determine the free radical formation mechanism. Alkyl R_1 will go through the Norrish Type I mechanism, while aryl R_1 will normally go through the Norrish Type II mechanism because CO-aryl bond energy is slightly higher and the available UV energy is insufficient to split this bond.

$$R_2 \!\!-\!\!\left[\bigcirc\right]\!\!-\!\!CO\!\!-\!\!R_1$$

Fig. (6).   Typical aryl ketone structure of free radical photoinitiators.

Fig. (7) shows the free radical generation mechanism of 2-hydroxy-2-methyl-1-phenyl-1-propanone *via* the Norrish Type I route [13]. This α-hydroxyketone photoinitiator will absorb suitable UV energy to become a very active excited triplet state that will split *via* Norrish Type I cleavage to two radicals: benzoyl radical and alkyl radical. The majority of commercialized photoinitiators are Type I photoinitiators. There are mainly four different chemical groups: α-hydroxyketone photoinitiators (HAPs), α-aminoketone photoinitiators (AAAPs), benzil ketal photoinitiators (BK), and phosphine oxide photoinitiators (POs).

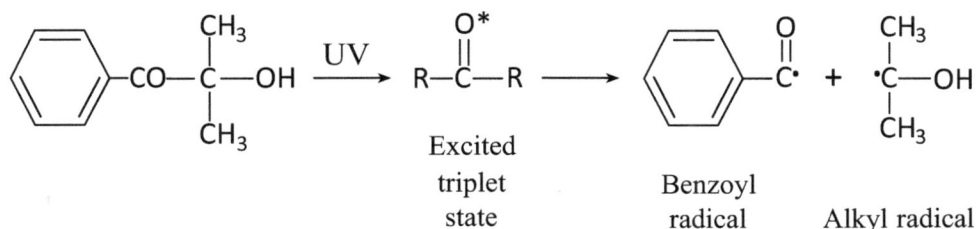

Fig. (7). Norrish Type I cleavage mechanism of α-hydroxyketone photoinitiator.

HAP photoinitiators have very strong absorption in the short UVC range around 230 – 270 nm and thus show very good surface curability. They also have some weak absorption at longer UVB ranges up to 360 nm and some depth curability. HAP photoinitiators are widely used and very effective. The chemical structure, appearance, CAS No., absorption peak, and key features of typical commercially available HAP photoinitiators are shown in Fig. (**8**). Absorption spectra of 1-hydroxy-cyclohexyl-phenylketone are shown in Fig. (**9**) [14].

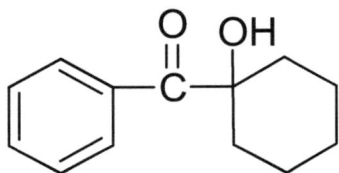

White powder
CAS. No.: 947-19-3
Absorption peak: 246, 280, 333 nm
Key features: low yellowing
 surface cure

1-hydroxy-cyclohexyl-phenylketone

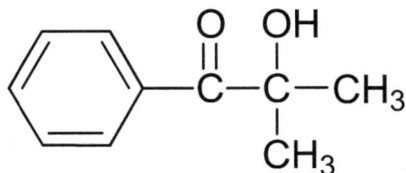

Amber liquid
CAS. No.: 7473-98-5
Absorption peak: 245, 280, 331 nm
Key features: liquid type
 low yellowing
 surface cure

2-hydroxy-2-methyl-1-phenyl-1-propanone

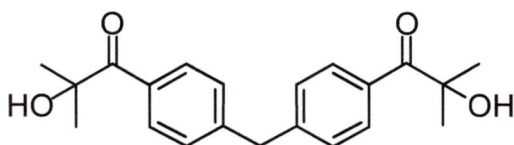

Light yellow powder
CAS. No.: 474510-57-1
Absorption peak: 260, 320 nm
Key features: high reactivity
low oxygen inhibition

2-hydroxy-1-(4-(4-(2-hydroxy-2-methyl-propyionyl)-benzyl)-

Phenyl)-2-methyl-propan-1-one

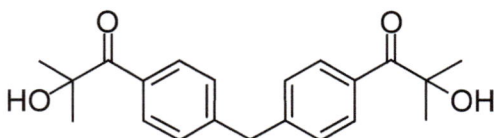

White powder
CAS. No.: 106797-53-9
Absorption peak: 276, 331 nm
Key features: low yellowing

2-hydroxy-1-(4-(2-hydroxyethoxy)phenyl)2-methyl-1-propanone

Fig. (8). Typical α-hydroxyketone photoinitiators.

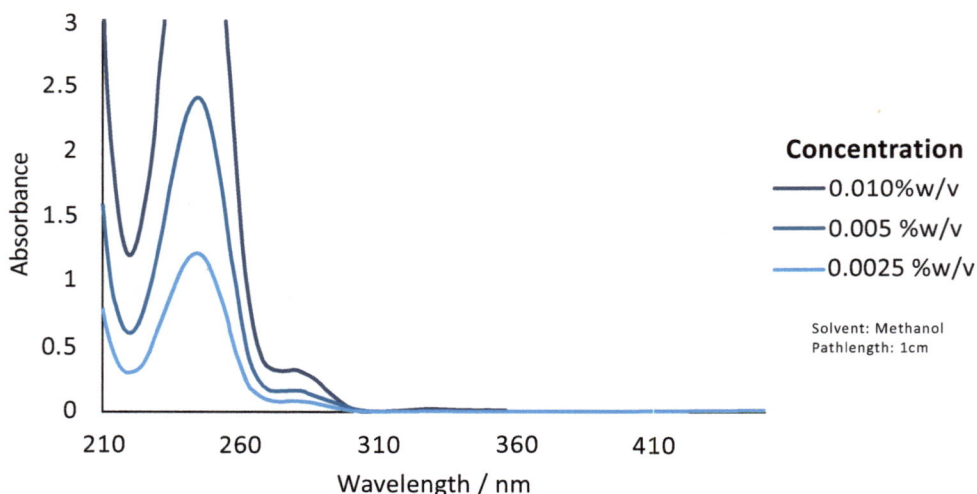

Fig. (9). Absorption spectra of 1-hydroxy-cyclohexyl-phenylketone.

AAAP photoinitiators have very strong absorption in the middle UVB range, around 280 – 350 nm. They are more reactive than HAP photoinitiators because of the amino group with higher electron density on the α-carbon, resulting in more efficient radical formation. AAAP photoinitiators are suitable for use in high-speed cure-required applications, but they tend to have yellowing issues.   The chemical structure, appearance, CAS No., absorption peak, and key features of typical commercially available AAAP photoinititors are shown in Fig. (**10**). Absorption spectra of 2-benzyl-2-dimethylamino-1-(4-morpholinophenyl)-butanone-1 are shown in Fig. (**11**) [15].

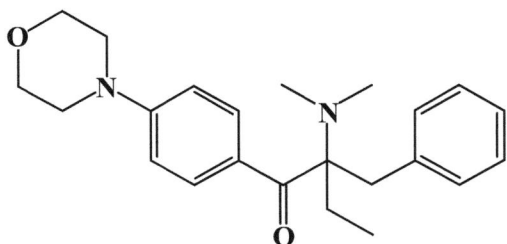

Light yellow powder
CAS. No.: 119313-12-1
Absorption peak: 233, 324 nm
Key features: highly efficient

2-benzyl-2-dimethyamino-1-(4-morpholinophenyl)-butanone-1

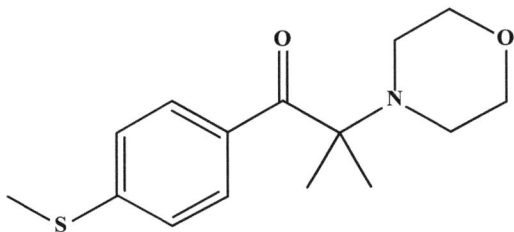

White powder
CAS. No.: 71868-10-5
Absorption peak: 232, 240, 307 nm
Key features: highly efficient

2-methyl-1-(4-(methylthio)phenyl)-2-morpholinopropan-1-one

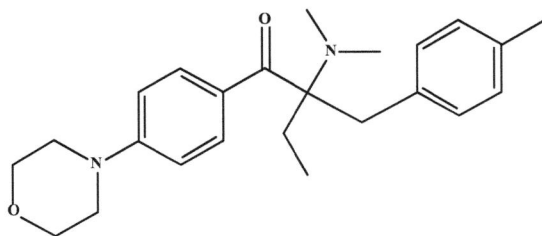

Light yellow powder
CAS. No.: 119344-86-4
Absorption peak: 237, 320 nm
Key features: highly efficient
good solubility

2-dimethylamino-2-(4-methyl-benzyl)-1-(4-morpholin-4-yl-phenyl)-butan-1-one

Fig. (10). Typical α-aminoketone photoinitiators.

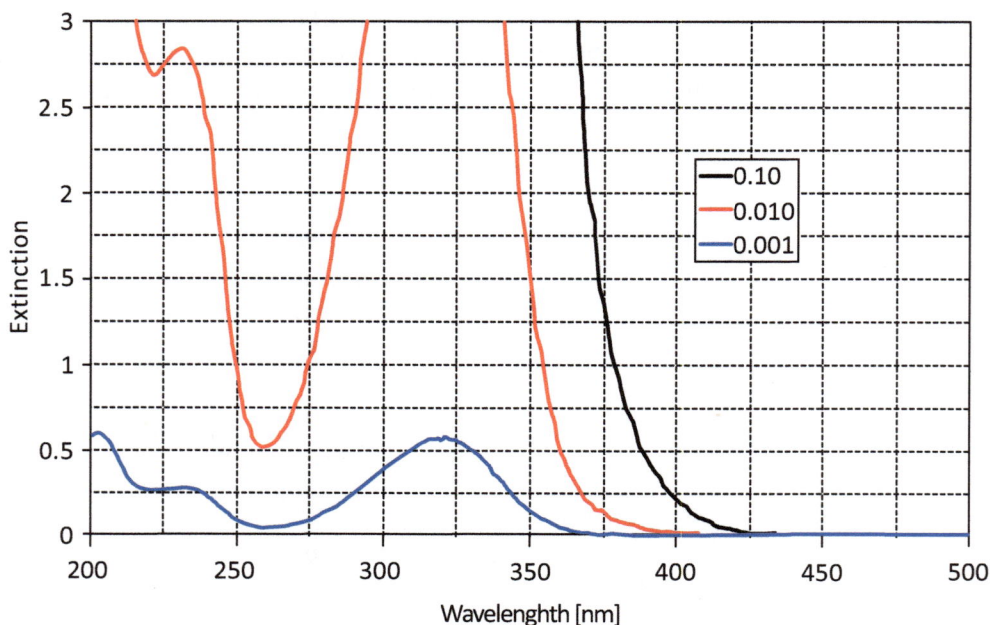

Fig. (11). Absorption spectra of a-aminoketone photoinitiators in acetonitrile.

BK photoinitiators have very strong absorption in the short UVC range and strong absorption in the UVA B range. 2,2 -Dimethoxy-1,2-diphenylethan-1-one (BDK) is one of the most widely used general-purpose photoinitiators. The chemical structure, appearance, CAS No., absorption peak, and key features of BDK photoinitiator are shown in Fig. (**12**). The absorption spectra of BDK are shown in Fig. (**13**) [16].

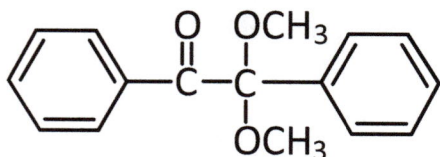

2,2 -Dimethoxy-1,2-diphenylethan-1-one

White powder
Cas. No.: 24650-42-8
Absorption peak: 254, 337 nm
Key features: highly efficient
                    surface cure

Fig. (12). Typical BK photonitiator.

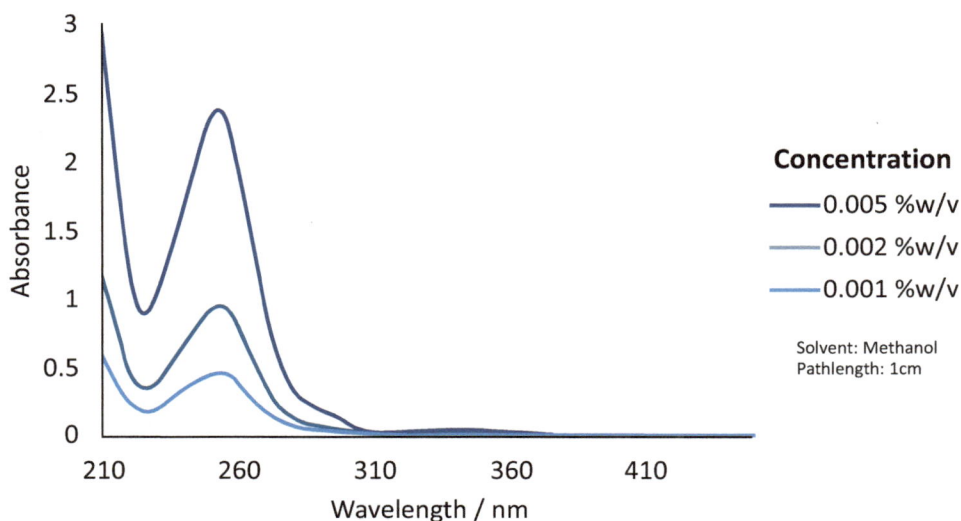

Fig. (13). Absorption spectra of BDK.

PO photoinitiators have absorption in the longer wavelength UVA and UV-Vis range around 350 – 420 nm. They are very effective depth cures and LED-UV light photoiniators. Although PO photoinitiators are yellow materials, the yellow color will be changed to white *via* UV or even daylight radiation. Surface curability can be improved by a combination use of HAP photoinitiators. The chemical structure, appearance, CAS No., absorption peak, and key features of typical commercially available PO photoinititors are shown in Fig. (**14**). Absorption spectra of 2,4,6-trimethylbenzoyl-diphenylphosphine oxide are shown in Fig. (**15**) [17].

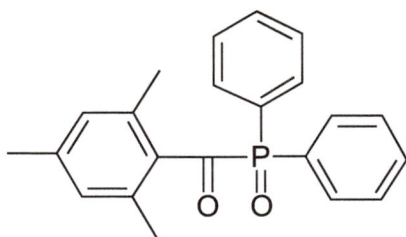

Yellow powder
CAS. No.: 75980-60-8
Absorption peak: 295, 380, 393 nm
Key features: low yellowing,
depth cure
long wavelength cure

2,4,6-trimethylbenzoyl-diphenylphosphine oxide

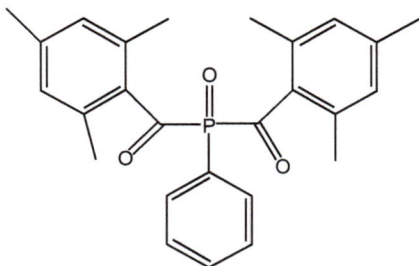

Yellow powder
CAS. No.: 162881-26-7
Absorption peak: 360, 365, 405 nm
Key features: depth cure
long wavelength cure

Bis(2,4,6-triethykbenzoyl)-phenyl phosphine oxide

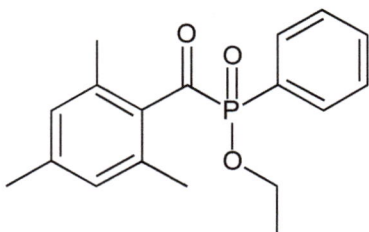

Yellow liquid
CAS. No.: 84434-11-7
Absorption peak: 242, 280, 370 nm
Key features: liquid type
depth cure
long wavelength cure

2,4,6-trimethylbenzoyl-diphenylphosphinate

Fig. (14). Typical acylphosphine oxide photoinitiators.

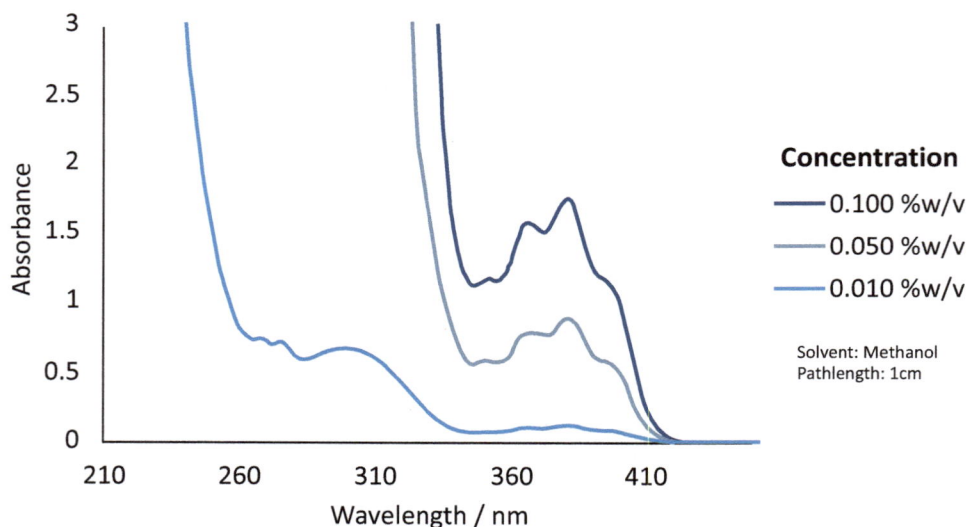

Concentration
— 0.100 %w/v
— 0.050 %w/v
— 0.010 %w/v

Solvent: Methanol
Pathlength: 1cm

Fig. (15). Absorption spectra of acylphosphine oxide photoinitiator TPO.

Fig. (**16**) shows the free radical generation mechanism of benzophenone *via* the Norrishi Type II route. Benzophenone will absorb suitable UV energy to become an excited triplet state that will need hydrogen donors such as tertiary amines to generate radicals: ketyl radical and donor radical. Benzophenones and substituted benzophenones are typical type II photoinitiators. Benzophenone has strong absorption in a short UVC wavelength range of around 230 – 260 nm. Benzophenone, with a combination of amine synergists, has very good surface curability and is a very cost-effective photoinitiator. The chemical structure, appearance, CAS No., absorption peak, and key features of typical benzophenone photoinitiator are shown in Fig. (**17**). Its absorption spectra are shown in Fig. (**18**) [18].

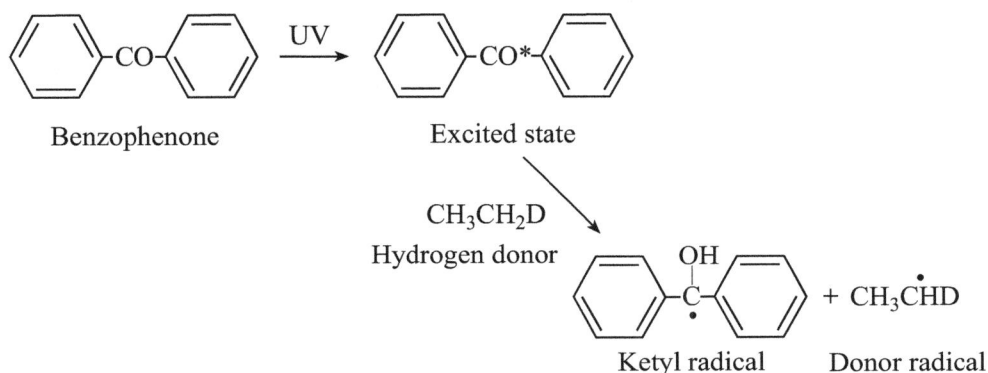

Fig. (16). Norrish Type II cleavage mechanism of benzophenone photoinitiator.

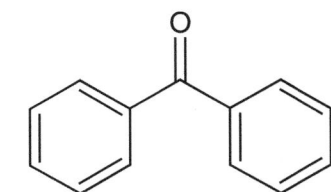

Benzophenone

White solid
Cas. No.: 119-61-9
Absorption peak: 252 nm
Key features: surface cure
cost effective

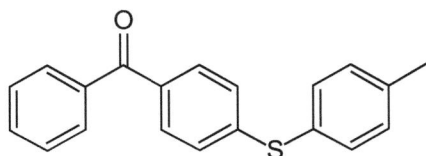

4-benzoyl-4'-methyldiphenyl sulfide

White solid
Cas. No.: 83846-85-9
Absorption peak: 246, 315 nm
Key features: low yellowing
depth cure

Fig. (17). Typical Type II benzophenone photoinitiator

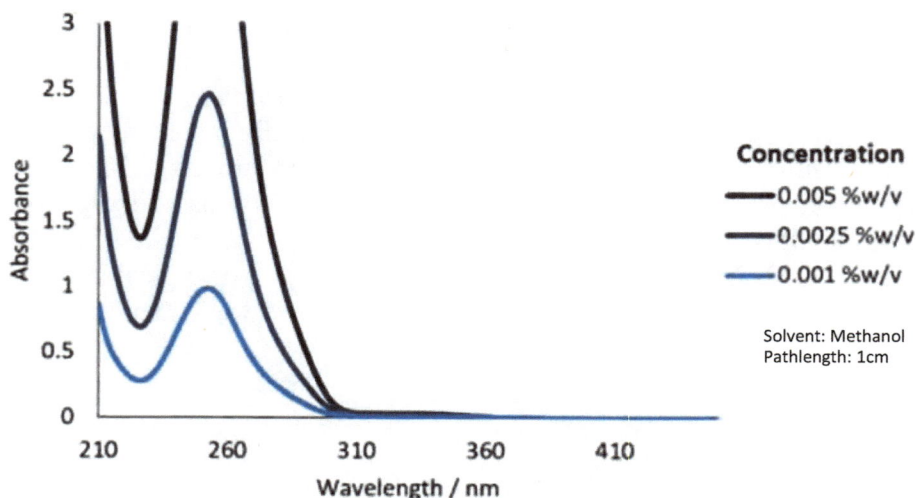

Fig. (18). Absorption spectra of benzophenone.

Polymerization Mechanism

Free radical photoinitiators absorb UV energy to generate free radicals that can initiate free radical polymerization of acrylate compositions, as illustrated in Fig. (**19**) [19]. The free radical polymerization mainly includes initiation, propagation, and termination steps. The initiation step is photosensitive, and the propagation and termination steps are thermally driven. The initiation is composed of UV light absorption, radical generation, and the reaction of a radical with an acrylate to form a reactive alkyl radical. The propagation is free radical polymerization of acrylate monomers, oligomers starting from the reactive alkyl radical to form an acrylic polymer network. Termination occurs when radicals interact or recombine together to form a neutral, stable chemical and stop chain polymerization. Free radical polymerization of acrylate will not occur in shadow areas where UV energy cannot be radiated, which is needed for free radical formation of photoinitiator.

Oxygen inhibits free radical polymerization. This phenomenon is called oxygen inhibition. Oxygen inhibition is always a big concern in free radical polymerization in atmosphere conditions where oxygen is contained, especially at surface area. Oxygen inhibits the polymerization of UV-cure acrylates in two ways: quenching the excited state of the photoinitiator and scavenging reactive radicals. Oxygen can quench the excited state of the photoinitiator and return it to the ground state, preventing radical formation as a result. Oxygen can scavenge reactive radicals *via* reaction with reactive radicals to form peroxy radicals, as shown in Fig. (20) [20]. The peroxy radicals are inactive to initiate polymerization but may terminate polymerization *via* radical-radical interaction.

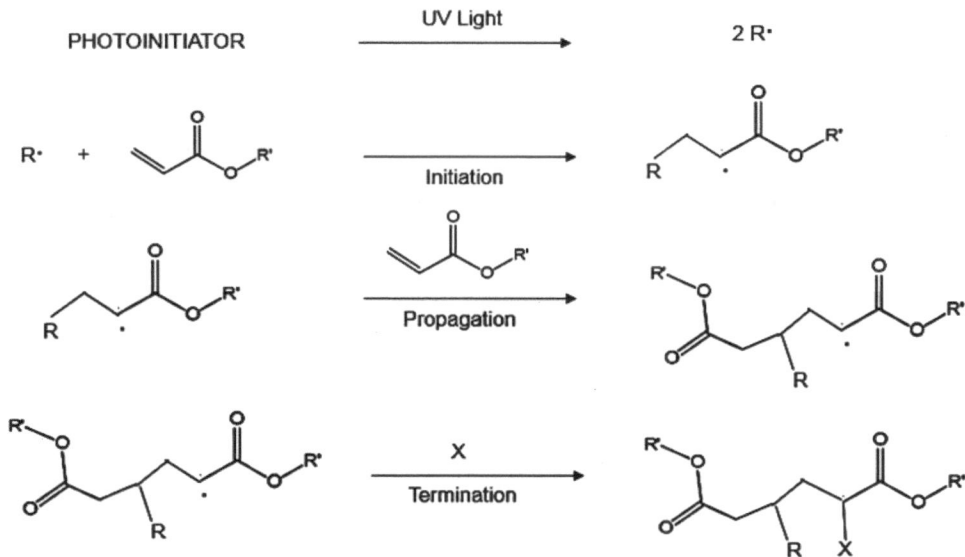

Fig. (19). Polymerization mechanism of UV cure acrylates.

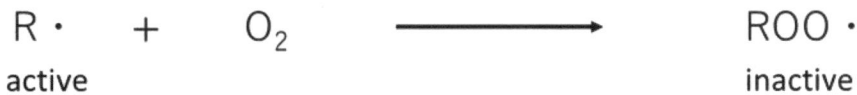

$$R\cdot \quad + \quad O_2 \quad \longrightarrow \quad ROO\cdot$$

active   inactive

Fig. (20). Inactive peroxy radical formation mechanism from active radical and oxygen.

Below are several practical methods used to reduce oxygen inhibition:

a. Nitrogen purging.

b. Combination with short wavelength UV source if available.

c. Higher UV intensity.

d. Higher photoinitiator content.

e. Incorporation with a small amount of wax

DUAL CURE HYBRID EPOXY ADHESIVES

UV and thermal dual cure hybrid epoxy adhesives are combined with UV cure acrylate composition and thermal cure epoxy composition [21-25]. They are mainly composed of acrylate monomer, epoxy resin, free radical photo-initiator, and epoxy curing agent. As illustrated in Fig. (21), these dual cure hybrid epoxy adhesives combine well the advantages of instant curability of UV acrylate composition with high-reliability performance of thermal cure epoxy part. The main components, polymerization mechanism, and key features of UV cationic epoxy adhesives, as well as dual cure hybrid epoxy adhesives, are compared with those of UV acrylate adhesives in Table 3. Dual cure hybrid epoxy adhesives have been successfully used in the electronics assembly industry, such as camera module assembly and display panel production applications where high mass production efficiency is required.

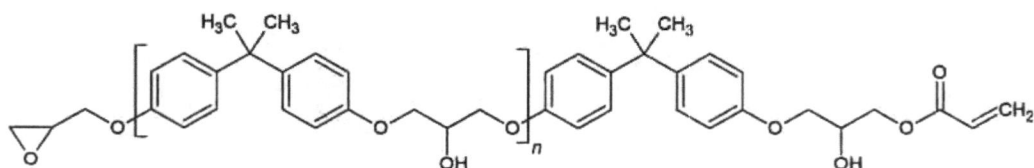

Epoxy Adhesives
Advantages
✓ No byproduct generation
✓ Low cure shrinkage
✓ Strong adhesion, high reliability performance
Limitations
• Rigid structure
• Long cure time

UV Cure Acrylate Adhesives
Advantages
✓ Fast cure
✓ Solvent free
✓ One-component, good stability
Limitations
• Oxygen inhibition
• Shadow cure problem
• High cure shrinkage

Fig. (21). Advantages and limitations of epoxy adhesives and UV cure acrylate adhesives.

Table 3. Comparison of UV cure acrylate, UV cationic epoxy, and dual-cure hybrid epoxy adhesives.

Adhesive type	UV cure Acrylate	UV Cure Cationic Epoxy	Dual Cure Hybrid Epoxy
Main components	Acrylate	Epoxy resin	Acrylate
	Photoinitiator	Cationic photoinitiator	Photoinitiator

(Table 3) cont.....

			Epoxy resin, Curing agent
Polymerization			
UV cure	Radical	Cationic	Radical
Thermal cure	N.A.	Cationic	Polyaddition, anionic
Key features			
Oxygen inhibition	Yes	No	Partial
Alkali inhibition	No	Yes	No
UV curability	High	Medium	High
Thermal cure	No need	Partially	Need
Shadow cure	No	Low	Yes
Cure shrinkage	High	Low	Medium
Adhesion	Moderate	Good	Good

Formulating Dual Cure Hybrid Epoxy Adhesives

Partially acrylated epoxy resin, epoxy resin, and acrylate oligomer are the adhesive bases used in the UV and thermal dual cure hybrid epoxy adhesives. As shown in Fig. (**22**), partially acrylated epoxy resin can cure with both epoxy resin components *via* the epoxy group and acrylate components *via* the acrylate group. The full cured composition will become a strong cross-linked hybrid polymer network that can bond to the adherend tightly. The most commonly used partially acrylated epoxy resin is partially acrylated bisphenol A epoxy resin, such as Ebecryl 3605, supplied by Allnex Corporation. Others are 3,4-epoxycyclohexylmethyl methacrylate and 3-ethyl-3-(methacryloyloxy) methyloxetane, whose chemical structure is shown in Fig. (**15**). Suitable amounts of epoxy resin and oligomer can be formulated as needed. Latent curing agents are typically selected as curing agents as dual cure hybrid epoxy adhesives are usually formulated in one-component formulation for easy handling.

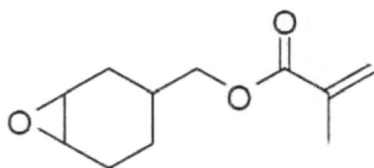

Viscosity: 9 mPas/25°C
CAS. No.: 82428-30-6
EEW: 225

3,4-Epoxycyclohexylmethyl methacrylate

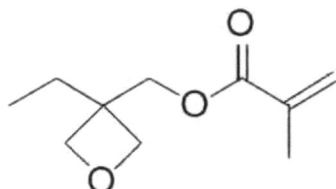

Viscosity: 5 mPas/25°C
CAS. No.: 37674-57-0
Molar mass: 184

3-Ethyl-3-(methacryloyloxy) methyloxetane

Fig. (22). 3,4-epoxycyclohexylmethyl methacrylate and 3-ethyl-3-(methacryloyloxy) methyl-oxetane.

Acrylate monomer plays a key role in the determination of its light cure behavior. Two or more acrylate monomers are often combined to achieve adhesive key requirements. A free radical photoinitiator is typically added with a small content in the range of 0.5 to 3%. The photoinitiator is selected based on light source and UV cure condition requirements. A combination of two or more photoinitiators might be needed, especially for long-wavelength light sources such as UV-LED lamp applications.

Fillers, tougheners, colorants, and other additives can also be used for UV and thermal dual-cure hybrid epoxy adhesives. Precautions need to be taken into consideration on their light transmittance. Cure depth will be significantly damaged in the case of poor light transmittance.

The function, component, main role, and typical content of UV and thermal dual cure hybrid epoxy composition are summarized in Table **4**.

Table 4. Dual cure hybrid epoxy adhesive composition.

Function	Component	Main Role	Content (%)
Primary	Epoxy resin/acrylate oligomer	Adhesive base	20 – 80

	Acrylate monomer	Handling and physical properties	10 – 30
	Curing agent	Curing epoxy components	0.5 – 40
	Radical photoinitiator	Curing acrylate components	0.5 - 3
Modifier	Filler	Property enhancement	0 – 70
	Toughener	Toughness enhancement	0 – 20
Additive	Colorant	Coloring	0 – 2
	Coupling agent	Adhesion promotion	0 – 2
	Thixotropic agent	Rheology control	0 – 5

Properties, Testing and Characterization

The four factors below should be taken into consideration when testing UV cure acrylate compositions:

- ✓ Lamp type.
- ✓ Light intensity.
- ✓ Temperature.
- ✓ Oxygen (atmosphere air).

Similar to UV cationic epoxy cases, the spectral output of the lamp type needs to be matched well with the absorbency property of the adhesive product's photoinitiators. Light intensity on the product is mainly a function of light source power, distance from the light source to the product, and the transmission property of the substrate when UV light is transmitted through. Light intensity and energy are measured with a UV radiometer for cure profile determination and cure process control. For UV and thermal dual cure hybrid epoxy adhesives, UV cure is primarily used to fix or temporarily hold on the adherends. Main physical properties and adhesion performance are normally measured after full cure with post-thermal cure conditions.

Curing properties are important for handling and cure condition determination. UV cure speed depends on both the product composition and actual light intensity. Any factor that reduces light intensity on the product will reduce cure speed. Depth of cure is enhanced by a longer wavelength as it can penetrate deeper into the product. Depth of cure is a function of product composition, wavelength, light intensity, and exposure time. Surface cure, in contrast, is enhanced by shorter, more energetic wavelengths. The cure behavior of thermal cure epoxy can be quantitatively analyzed by differential scanning calorimetry (DSC) to measure changes in enthalpy generated during thermal curing. FT-IR is often used to analyze cure conversion or degree of cure of acrylate and epoxy compositions by measuring changes in the characteristic absorption peak area of acrylate, epoxy group, and oxetane group during and after light exposure and thermal heating. Characteristic absorption of acrylate, epoxy, and oxetane groups is summarized in Table **5** [26].

Table 5. FT-IR characteristic absorption peak of acrylate, epoxy, and oxetane groups.

Functional Group	Characteristic Absorption
Acrylate	
$CH_2=CH-COO$'s stretching vibration	1635 & 1619 cm^{-1}
$CH_2=CH-COO$'s in-plane bending vibration	1405 cm^{-1}
$CH_2=CH-COO$'s out-of-plane bending vibration	810 cm^{-1}
Methacrylate	
$CH_2=C(CH_3)-COO$'s stretching vibration	1639 cm^{-1}
$CH_2=C(CH_3)-COO$'s in-plane bending vibration	1403 cm^{-1}
$CH_2=C(CH_3)-COO$'s out-of-plane bending vibration	814 cm^{-1}
Epoxy group, glycidyl ether type	
Epoxy group's antisymmetric stretching vibration	916 cm^{-1}
C-O-C's antisymmetric stretching vibration	1108 cm^{-1}

(Table 5) cont.....

Epoxy group, cycloaliphatic type	
Epoxy group's antisymmetric stretching vibration	789 cm^{-1}
Epoxy group's symmetric stretching vibration	746 cm^{-1}
C-O-C's antisymmetric stretching vibration	1088 cm^{-1}
Oxetane group, oxetanes	
Oxcetane group's antisymmetric stretching vibration	995 cm^{-1}
C-O-C's antisymmetric stretching vibration	1100 cm^{-1}

APPLICATIONS

LCD ODF Main Sealing

The successful development and industrialization of the so-called ODF (One Drop Fill) process for large-size LCD (liquid crystal display) panel production was an important technological revolution in the early 2000s that had a big impact on our daily lives. The development of LCD ODF main sealant, a UV and thermal dual cure hybrid epoxy adhesive used to bond two glass substrates and seal liquid crystal material between them, played a key role in its mass production [27, 28]. LCD ODF panel assembly process is illustrated in Fig. (**23**) [29].

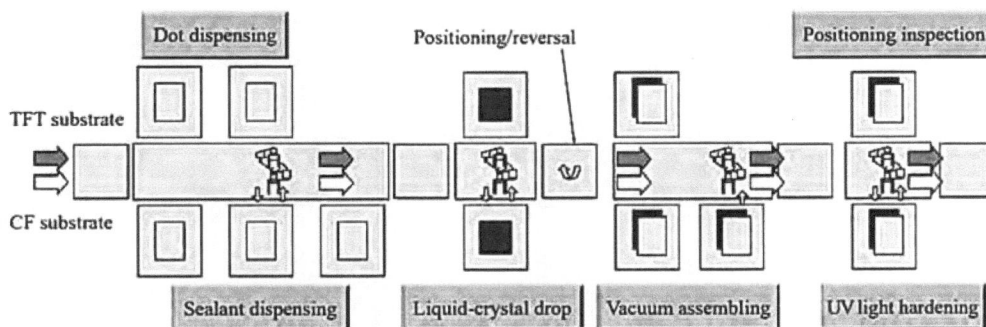

Fig. (23). LCD ODF assembly process.

We invented an initiator-free UV and thermal dual cure hybrid epoxy adhesive by combining with bismaleimides that shows much better compatibility with liquid crystal material and high-reliability performance as a new LCD ODF main sealant [30-36]. As shown in Fig. (24), the invented dual cure hybrid epoxy adhesive is mainly composed of specially designed liquid bismaleimide, acrylate monomer, partially acrylated bisphenol A epoxy resin, bisphenol A epoxy resin, and latent curing agent. Its UV and thermal cure behavior was investigated by FT-IR spectroscopy and FT-Raman spectroscopy. FT-IR analysis results are summarized in Table 6. The initiator-free dual cure hybrid epoxy adhesive showed satisfactory UV curability, good shadow curability, and high adhesion performance. LCD ODF main sealant developed based on this invention has been successfully used in actual LCD panel production.

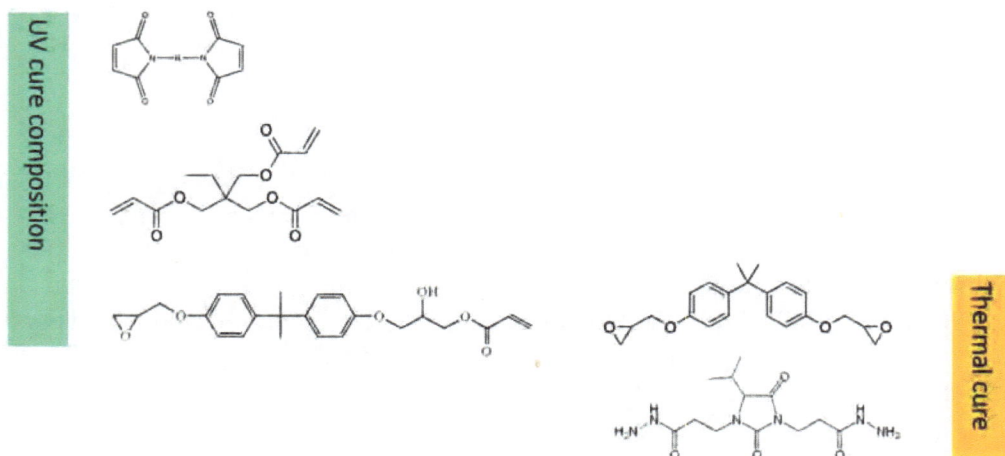

Fig. (24). Raw materials used for initiator-free dual cure epoxy investigation.

Table 6. Conversion rates of C=C group and epoxy groups measured by FT-IR.

Cure Condition	C=C Conversion Rate (%)		Epoxy Conversion Rate (%)
	Acrylate	Bismaleimide	
UV cure only 100 mW/cm² x 30 sec	62	95	0
UV+ thermal cure	100	96	85

(Table 6) cont.....

100 mW/cm² x 30 sec + 120°C x 60 min			
Thermal cure only 120°C x 60 min	67	95	69

Camera Module Assembly

Mobile phone camera technology is another important technological revolution that has also had a big impact on our daily lives in recent decades. Epoxy adhesives have been used as the main adhesive to bond lens holders with image sensor substrates in camera module assembly from the very beginning of their commercial production because of the satisfactory adhesion and reliability performance of epoxy adhesive [37-40]. In recent years, UV and thermal dual cure hybrid epoxy adhesives have become more and more popular due to high production efficiency requirements. As illustrated in Fig. (**25**) [41], the main steps for the use of dual cure hybrid epoxy adhesive in this process are: 1) dispensing dual cure hybrid epoxy adhesive on image sensor substrate, 2) position, 3) alignment and assembly, 4) UV cure fixing, 5) thermal post cure, and 6) assembly completion.

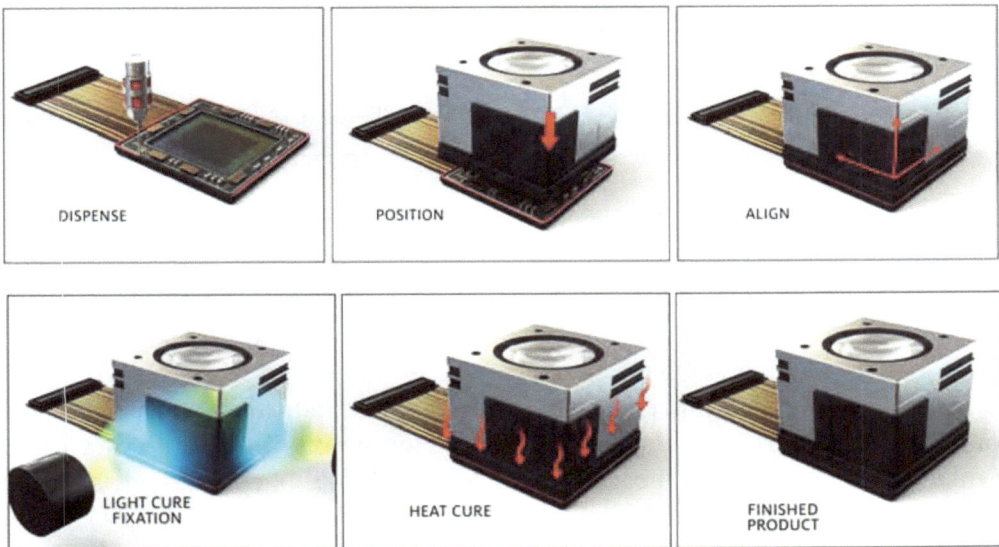

Fig. (25). Typical dual cure camera module assembly process.

Loctite 3217 was one of the first dual cure hybrid epoxy adhesives especially designed for mobile phone camera assembly. Product description and typical properties are summarized in Table 7 [42].

Table 7. Product description and typical properties of LOCTITE 3217.

Appearance	Amber Viscous Liquid
Chemical type	One-component epoxy
Cure	Ultraviolet (UV) light + thermal cure
Viscosity	37600 mPa.s/25°C
Density	1.2
Pot life	2 weeks @ 25°C
Hardness*, Shore D	86
Volume shrinkage	5.6 %
Tg, measured by TMA	82°C
UV tack-free time	1 second @ 100 mW/cm², medium pressure mercury lamp
Recommend cure condition	30 minutes @ 60°C
	20 minutes @ 80°C
Storage condition	Refrigeratory (2 to 8°C)

*Cure condition: 80°C x 20 minutes.

CONCLUSION

Dual cure hybrid epoxy adhesives are primarily composed of partially acrylated epoxy resin, acrylate oligomer, acrylate monomer, free radical photoinitiator, epoxy resin, and latent curing agent. Partially acrylated epoxy resin, epoxy resin, and acrylate oligomer are the adhesive bases. A free radical photoinitiator is selected based on the light source and UV cure condition requirements. Lamp type, light intensity, and oxygen (air atmosphere) need to be taken into consideration when testing and using UV acrylate adhesives. The dual cure hybrid epoxy adhesives combine very well the advantages of instant curability of UV acrylate composition with high-reliability performance of thermal cure epoxy part. Dual cure hybrid epoxy technology has been very successfully used in general structural bonding, camera module assembly, and LCD panel bonding applications.

REFERENCES

[1]   *Radiation Curing Science and Technology*; Springer Science: New York, **1992**.

[2]   Glockner, P.; Jung, T.; Struck, S. *Studer, K. Radiation Curing Coatings and Printing Inks*; Vincentz Network: Hannover, Germany, **2008**.

[3]   Vitale, A.; Trsusiano, G.; Bongiovanni, R. UV-curing of adhesives: a critical review. In: *Progress in Adhesion and Adhesives*; Mittal, K.L., Ed.; Scrivener: Beverly, MA, **2018**, Vol. 3, pp. 101-154.
        http://dx.doi.org/10.1002/9781119526445.ch4

[4]   Velankar, S.; Pazos, J.; Cooper, S.L. High-performance UV-curable urethane acrylates *via* deblocking chemistry. *J. Appl. Polym. Sci.,* **1996**, *62*(9), 1361-1376.
        http://dx.doi.org/10.1002/(SICI)1097-4628(19961128)62:9<1361::AID-APP6>3.0.CO;2-F

[5]   Dekker, C. UV-radiation curing of adhesives. In: *Handbook of Adhesive and Surface Preparation*; Ebnesajjad, S., Ed.; Elsevier: New York, **2011**, pp. 221-243.
        http://dx.doi.org/10.1016/B978-1-4377-4461-3.10010-0

[6]   Haddon, M.R.; Smith, T.J. The chemistry and applications of UV-cured adhesives. *Int. J. Adhes. Adhes.,* **1991**, *11*(3), 183-186.
        http://dx.doi.org/10.1016/0143-7496(91)90021-9

[7]   Javadi, A.; Mehr, H.S.; Sobani, M.; Soucek, M.D. Cure-on-command technology: A review of the current state of the art. *Prog. Org. Coat.,* **2016**, *100*, 2-31.
        http://dx.doi.org/10.1016/j.porgcoat.2016.02.014

[8]   Schwalm, R. *UV Coatings Basics, Recent Developments and New Applications*; Elsevier: Amsterdam, **2007**.

[9]   Moeck, A.; Bianchi, R.; Petry, V.; Weder, R.; Helsby, D. Advances in epoxy resins for coatings. *J. Coat. Technol. Res.* **2014**, 11 (4), 423-431.

[10]   Chen, C.; Sun, D.; Kanari, M.; Lu, D. Study on the cure behavior of a novel photocurable material using UPLC-Q-TOF-MS. *Adv. Mater.,* **2020**, *9*(1), 8-14.
        http://dx.doi.org/10.11648/j.am.20200901.12

[11]   Xiong, T.; Zhang, Y. Synthesis and properties of polyurethane acrylate oligomer based on polycaprolactone diol. *e-Polymers,* **2022**, *22*, 147-155.

[12]   Oprea, S.; Vlad, S.; Stanciu, A.; Macoveanu, M. Epoxy urethane acrylate. *Eur. Polym. J.* **2000**, 36, 373-378.
        http://dx.doi.org/10.1016/S0014-3057(99)00077-4

[13]   Green, W.A. *Industrial phtoinitiator*; CRC Press: New York, **2010**, p. 19.
        http://dx.doi.org/10.1201/9781439827468

[14] *Speedcure 84;* Arkema Corporation. https://emea.sartomer.arkema.com/en/product-finders/product/f/sartomer_Lambson/p/speedcure-84/ [Accessed on December 20, 2023].

[15] *Omnirad 369;* IGM Resins. https://www.igmresins.com/en/product/omnirad_369 [Accessed on December 20, 2023].

[16] Speedcure BKL, Arkema Corporation. https://emea.sartomer.arkema.com/en/product-finders/product/f/sartomer_Lambson/p/speedcure-bkl/ [Accessed on December 20, 2023].

[17] *Speedcure TPO;* Arkema Corporation. https://americas.sartomer.arkema.com/en/-product-finders/product/f/sartomer_Lambson/p/speedcure-tpo/ [Accessed on December 20, 2023].

[18] *Speedcure BP;* Arkema Corporation. https://asia.sartomer.arkema.com/en/product-finders/product/f/sartomer_Lambson/p/speedcure-bp/ [Accessed on December 20, 2023].

[19] Bednarczyk, P.; Mozelewska, K.; Czech, Z. Influence of the UV crosslinking method on the properties of acrylic adhesive. *Int. J. Adhes. Adhes.,* **2020**, *102*, 102652.
http://dx.doi.org/10.1016/j.ijadhadh.2020.102652

[20] Zhang, W.; Shentu, B.; Weng, Z. Preparation and properties of heat and ultraviolet-induced bonding and debonding epoxy/epoxy acrylate adhesives. *J. Appl. Polym. Sci.,* **2018**, *135*(26), 46435.
http://dx.doi.org/10.1002/app.46435

[21] Lee, J.H. Highly adhesive and sustainable UV/heat dual-curable adhesives embedded with reactive core-shell polymer nanoparticles for super-narrow bezel display. *Materials (Basel),* **2020**, *13*(16), 3492.
http://dx.doi.org/10.3390/ma13163492 PMID: 32784686

[22] Yildiz, Z.; Gungor, A.; Onen, A.; Usta, I. Synthesis and characterization of dual-curable epoxyacrylates for polyester cord/rubber applications. *J. Ind. Text.,* **2016**, *46*(2), 596-610.
http://dx.doi.org/10.1177/1528083715594980

[23] Park, C.H.; Lee, S.W.; Park, J.W.; Kim, H.J. Preparation and characterization of dual curable adhesives containing epoxy and acrylate functionalities. *React. Funct. Polym.,* **2013**, *73*(4), 641-646.
http://dx.doi.org/10.1016/j.reactfunctpolym.2013.01.012

[24] Xiao, M.; He, Y.; Nie, J. Novel bisphenol A epoxide-acrylate hybrid oligomer and its photopolymerization. *Des. Monomers Polym.,* **2008**, *11*(4), 383-394.
http://dx.doi.org/10.1163/156855508X332522

[25] Park, Y.J.; Lim, D.H.; Kim, H.J.; Park, D.S.; Sung, I.K. UV- and thermal-curing behaviors of dual-curable adhesives based on epoxy acrylate oligomers. *Int. J. Adhes. Adhes.,* **2009**, *29*(7), 710-717.
http://dx.doi.org/10.1016/j.ijadhadh.2009.02.001

[26] Tanaka, H. Cure behavior analysis for UV cure resins. In: *Optimization on UV cure process*; Fukushima, H., Ed.; Science and Technology Publishing: Tokyo, **2008**, pp. 50-64.

[27]   Xiao, A.; Song, S.; Zhang, W.; Chen, X.; Min, T. Study of uncured sealant contaminating liquid crystal in one drop filling process of thin film transistor liquid crystal display. *SID Int. Symp. Dig. Tec.,* **2012**, pp. 1134-1136.

[28]   Chen, C.; Iida, K. Adhesive for flat-panel display manufacture In: *In: Ullmann's Encyclopedia of Industrial Chemistry, Adhesives, 2. Applications*; Henkel Expert Team, Ed.; Wiley – VCH: Weinheim, Germany, **2012**, pp. 519-520.

[29]   Hirai, A.; Abe, I.; Mitsumoto, M.; Ishida, S. One drop filling for liquid crystal display panel produced from larger-sized mother glass. *Hitachi Rev.,* **2008**, *57*, 144-148.

[30]   Chen, C. Sealing agent for liquid crystal dropping technology and method of manufacturing liquid crystal display. *Japanese Patent 5592081,* **2014**.

[31]   Chen, C. Sealant composition. *U.S. Patent 10108029,* **2018**.

[32]   Chen, C. Method of manufacturing liquid crystal display panel. *Japanese Patent 5592111,* **2014**.

[33]   Chen, C. Bonding method and adhesive resin composition. *Chinese Patent 101553545,* **2012**.

[34]   Zhou, J.; Wang, T.; Li, Q.; Chen, F. Curable resin composition for sealing liquid crystal. *Chinese Patent Application 109416487,* **2019**.

[35]   Chen, C.F.; Iwasaki, S.; Kanari, M.; Li, B.; Wang, C.; Lu, D.Q. High performance UV and thermal cure hybrid epoxy adhesive. *IOP Conf. Series Mater. Sci. Eng.,* **2017**, *213*, 012032.
        http://dx.doi.org/10.1088/1757-899X/213/1/012032

[36]   Chen, C. One-component instant bonding epoxy adhesives, in: *Proceedings of IUPAC – MACRO 2020+ The 48th World Polymer Congress*, 2OS9-2, Jeju, Korea, May 16 – 20, **2021**.

[37]   Henkel Corporation. New adhesive materials for image sensor housings now available from Henkel, *Henkel Corporation.* https://www.circuitnet.com/pdf/pr_86_new_image_sensor_housing_materials.pdf [accessed on Dec 20, 2023].

[38]   Henkel Corporation. Adhesives are a driving force in the automotive mobility space, *Henkel Corporation.* https://www.henkel-adhesives.com/us/en/insights/all-insights/blog/advanced-driver-assistance-systems-adas.html [accessed on Dec 20, 2023].

[39]   Bitzer, K.; By, A. Active alignment for cameras in mobile devices and automotive applications. Proceedings of the 2010 *IEEE Electronic Packaging Technology Conference* **2010**.
        http://dx.doi.org/10.1109/EPTC.2010.5702644

[40]   Henkel Corporation. Advanced camera modules require dual function adhesives, *Henkel Corporation.* https://dm.henkel-dam.com/is/content/henkel/whitepaper-1015-camera-modules-and-dual-cure-adhesives [accessed on Dec 20, 2023].

[41]   Henkel Corporation. Material solution for compact camera modules, *Henkel Corporation.* https://dam.com/is/content/henkel/lt-7014-brochure-materials-for-compact-camera-modules [accessed on Dec20, 2023].

[42]   Henkel Corporation. *Loctite* 3217, https://datasheets.tdx.henkel.com/LOCTITE-3217-en_GL.pdf [accessed on Dec 20, 2023].

CHAPTER 4

Snap Thermal Cure Epoxy Technology

Abstract: Snap thermal cure epoxy adhesives are typically one-component epoxy composition, comprising epoxy resin, a new type latent curing agent, and an accelerator with modifiers and additives. Snap thermal cure adhesives can cure very fast at certain elevated temperature conditions and have been very successfully used in semiconductor packaging and electronics module assembly applications.

Keywords: Accelerator, Anisotropic conductive film (ACF), Dicyandiamide, Imidazole, Non-conductive paste (NCP), Thermal cationic initiator.

LATENT CURING AGENTS

Snap thermal cure epoxy technology is mainly based on one-component epoxy adhesive, typically formulated with epoxy resin, latent curing agent, and accelerator with various modifiers and additives. New type latent curing agents, such as modified imidazole type, modified polyamine type, and thermal cationic initiator, have been developed and commercialized in recent decades. One-component thermal cure epoxy adhesives incorporated with these new type curing agents can cure very fast at certain elevated temperature conditions and have been successfully used in semiconductor packaging and electronics assembly applications.

One-component epoxy adhesives are prepared by selecting latent curing agents. All ingredients, including epoxy resin and curing agent, are mixed thoroughly in advance. No additional pre-mixing process in actual use is required. One-component adhesives can be handled easily and are suitable for automatic dispensing systems because of their long enough pot life.

With the selection of suitable latent curing agents, various one-component thermal cure epoxy adhesives have been developed and supplied by major epoxy adhesive suppliers for different applications. Typical commercial latent curing agents are summarized in Table **1** [1].

Imidazoles

Imidazole compounds react with an epoxy group to form anionic species, which can initiate the polymerization of epoxy resin, as illustrated in Fig. (**1**) [2]. The additional amount of imidazole is quite small, ranging from 1 to 8 phr based on standard bisphenol A epoxy resin (EEW=190). Pot life is longer than normal amine-type curing agent, ranging from a few hours to several months at room temperature, depending on its structure. A thermal cure at elevated temperature is needed to achieve a full cure, but cure time becomes significantly shorter. Cured epoxy resin based on imidazole shows good adhesion and high glass transition temperature. Like tertiary amine, imidazoles are also used as an accelerator for other epoxy-curing agents such as anhydrates, phenols, and dicyandiamide.

Table 1. Typical commercial latent curing agents.

Latent curing agent	Latency Mechanism	Curing Agent State	Typical Curing Temperature
Imidazoles		Liquid/solid	$\geqq 80°C$
DICY	Chemical blocking/physical	Solid	$\geqq 150°C$
Dihydrazine	separation		$\geqq 120°C$
Modified imidazole	Physical separation	Fine powder	$\geqq 80°C$
Modified polyamine			$\geqq 80°C$
Onium salts	Chemical blocking	Solid	$\geqq 80°C$
Amine-BF$_3$ complex		Liquid	$\geqq 130°C$

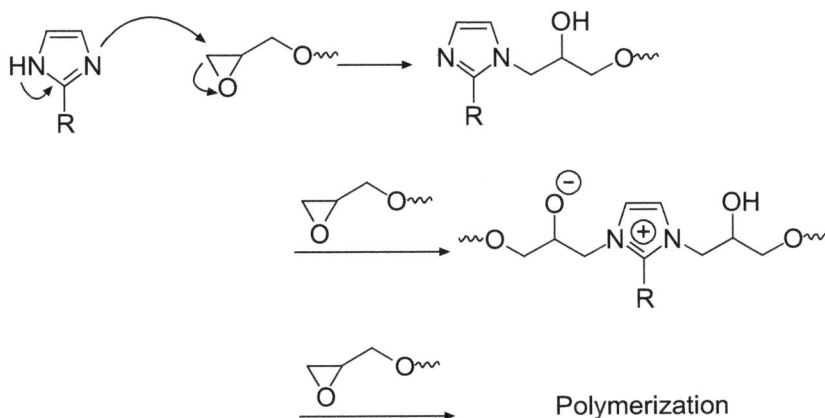

Fig. (1). Anionic polymerization of epoxy resin initiated by imidazole compound.

The chemical structure and physical properties of typical imidazole-type curing agents are shown in Fig. (2). 2-methyl imidazole and 2-ethyl-4-methyl imidazole are most widely used. 2-methyl imidazole is a white solid. The pot life of formulated epoxy resin is quite short, around 5 hours at room temperature. It has very good thermal curability. 2-methyl imidazole is often used as raw material for the synthesis of modified imidazole curing agents. 2-Ethyl-4-methyl imidazole is a pale-yellow liquid and has very good handling properties. The pot life of the formulated epoxy resin is around 8 hours at room temperature. It has very good thermal curability with a gel time of around 3 minutes at 120°C and around 1 minute at 150°C. Table **2** shows the cure behavior and physical properties of standard bisphenol A epoxy resin and 2-ethyl-4-methyl imidazole composition. 2-phenyl imidazole is a white solid [3]. The pot life of the formulated epoxy resin is around 20 hours at room temperature. It needs some higher temperature cure. Gel time is 17 minutes at 100°C. 1-Cyanoethyl-2-ethyl-4-methylimidazole has a longer pot life, around 4 days at room temperature, but the cure temperature needs to be slightly higher than 2-ethyl-4-methyl imidazole.

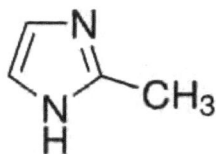

White solid
CAS. No.: 693-98-1
Molar mass: 82
Melting point: 142°C

2-Methyl imidazole

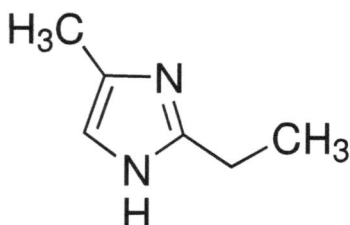

Ple yellow liquid
CAS. No.: 931-36-2
Molar mass: 110
Melting point: 47-54°C

2-Ethyl-4-methyl imidazole

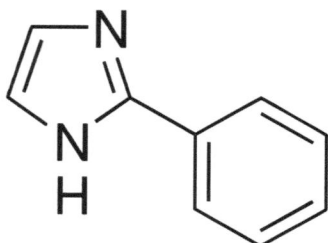

White solid
CAS. No.: 970-90-2
Molar mass: 144
Melting point: 142-148°C

2-Phenyl imidazole

CH₃

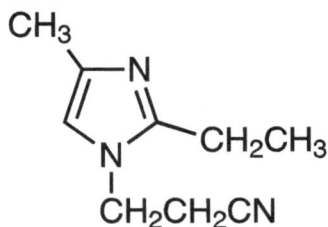

Light yellow liquid
CAS. No.: 568591-00-4
Molar mass: 163
Melting point: 11°C

1-Cyanoethyl-2-ethyl-4-methylimidazole

Fig. (2). Chemical structure and physical properties of typical imidazole catalysts.

Table 2. Epoxy composition cured with 2-ethyl-4-methyl imidazole.

Composition	
Standard Bisphenol A Epoxy Resin (EEW=190)	**100**
2-ethyl-4-methyl imidazole	**3**
Pot life, @25°C	8 hours
Gel time	215 seconds @120°C
	63 seconds @150°C
Tg*, measured by TMA method	162 °C
Hardness, shore D	85
Bending strength, MPa @25°C	79.4
Bending modulus, MPa @25°C	2569

*cure condition: 60°C x 4 hours + 150°C x 4 hours.

Dicyandiamide

Dicyandiamide, often called DICY, is a solid chemical with a melting point of 208°C. The chemical structure and key physical properties of DICY are shown in Fig. (3). DICY incorporated epoxy composition is very stable and has a relatively long pot life, usually more than 6 months at room temperature. DICY-cured epoxy resin possesses very high adhesion strength and high Tg, making it suitable for use in applications requiring high performance. Its cure temperature with epoxy resin needs to be quite high, around 180°C for full cure. The cure reaction between DICY and epoxy is very complicated with a combination of polyaddition of its amine active hydrogen with an epoxy group, its cyano group with an epoxy group, its cyano group with a hydroxy group, and anionic polymerization of an epoxy group initiated *via* tertiary amine. The chemical reactions of DICY and epoxy resin are illustrated in Fig. (4) [4]. Commercial grade DICY is supplied as a fine powder. The particle size will affect its curability and latency. A typical additional amount of DICY in standard bisphenol A epoxy resin (EEW=190) is around 4 to 10 phr.

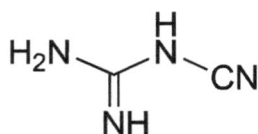

White, crystalline particles
CAS. No.: 461-58-5
Molar mass: 84
Melting point: 209.5°C

Fig. (3). Chemical structure and physical properties of dicyandiamide (DICY).

(1) Epoxy / primary amine (2) Epoxy / secondary amine reaction

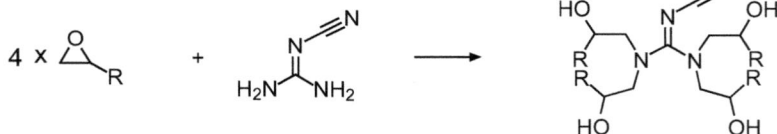

(3) Cyano / hydroxyl reaction

(4) Cyano / epoxy reaction

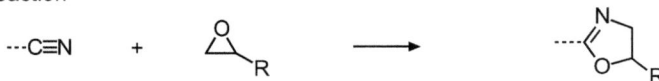

(5) Anionic polymerization

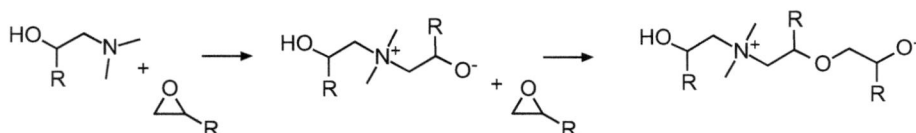

Fig. (4). Cure reactions of DICY and epoxy resin.

Accelerators such as substituted urea, tertiary amine, and imidazole compounds are often combined to reduce the cure temperature of DICY. Substituted urea has been widely used as an accelerator for DICY because of its fast curability, long pot life, and good adhesion performance. Substituted urea will decompose at elevated temperatures to dimethyl amine and isocyanate compounds. Dimethylamine can accelerate the cure reaction of epoxy resin and DICY. The chemical structure and key physical properties of typical substituted urea accelerators are shown in Fig. (5). Phenyl dimethyl urea and 2,4- toluene bis-dimethyl urea have the fastest curability. A typical additional amount of substituted urea is between 1 to 5 phr based on standard bisphenol A epoxy resin. With the selection of a suitable accelerator, the cure temperature can be significantly lowered to 120°C or even 110°C.

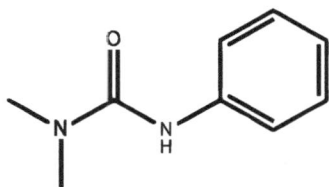

White powder
CAS. No.: 101-42-8
Molar mass: 164
Melting point: 132°C

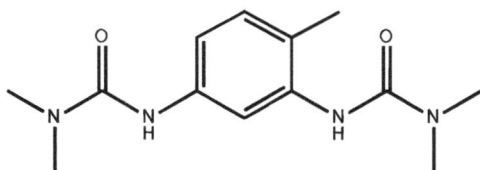

Phenyl dimethyl urea

White powder
CAS. No.: 17526-94-2
Molar mass: 264
Melting point:°C

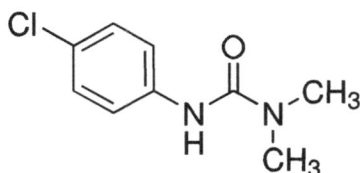

2,4—Toluene bis-dimethyl urea

White powder
CAS. No.: 150-68-5
Molar mass: 198
Melting point: 173°C

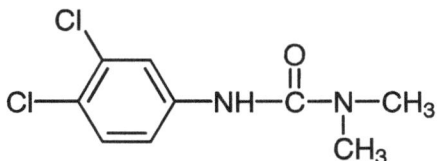

N-(4-chlorophenyl)N,N-dimethyl urea

White powder
CAS. No.: 330-54-1
Molar mass: 233
Melting point: 157°C

3-(3,4-dichlorophenyl)-1,1-dimethylurea (DCMU)

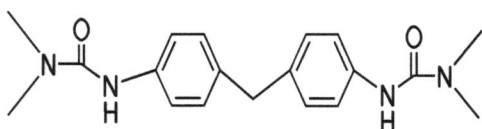

White powder
CAS. No.: 10097-09-3
Molar mass: 340
Melting point: 232°C

4,4'-Methylene bis phenyl dimethyl urea

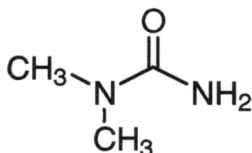

White powder
CAS. No.: 598-94-7
Molar mass: 88
Melting point: 182°C

1,1,-Dimethyl urea

Fig. (5). Chemical structure and physical properties of substituted urea accelerators.

Dihydrazides

Dihydrazides can react with an epoxy group by its primary amine *via* a polyaddition mechanism almost equivalently at elevated temperature conditions, as illustrated in Fig. (**6**) [5]. The chemical structure and physical properties of typical dihydrazide type curing agents are shown in Fig. (**7**). Dihydrides are usually solid chemicals. Cure temperature and storage stability of their epoxy compositions relate closely to the melting temperature. Accelerators such as imidazoles and ureas can be combined to improve their cure behavior. Cured epoxy compositions have good adhesion performance [6-8].

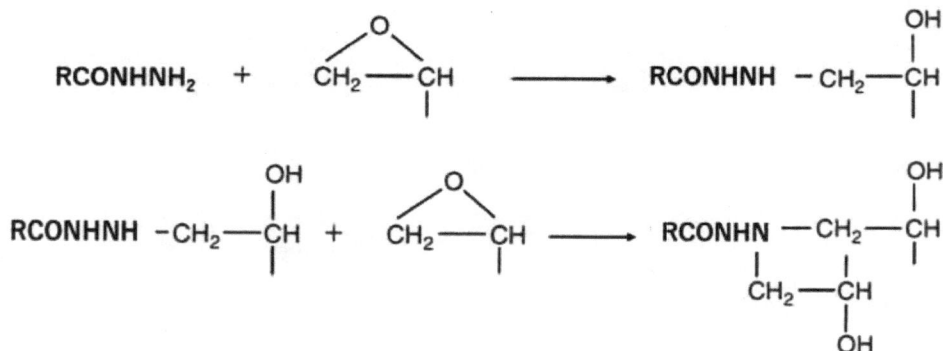

Fig. (6). Polyaddition reaction of epoxy resin with hydrazide compound.

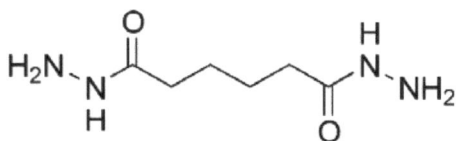

Adipic dihydrazide (ADH)

White powder
CAS. No.: 1071-73-8
Molar mass: 175
Melting point: 180°C

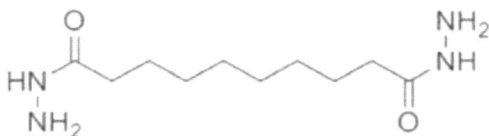

Sebacic dihydrazide (SDH)

Crystalline powder
CAS. No.: 125-83-7
Molar mass: 230
Melting point: 186°C

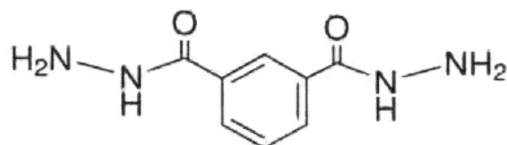

Isophthalic dihydrazide (IDH)

Crystalline powder
CAS. No.: 2760-98-7
Molar mass: 194
Melting point: 227°C

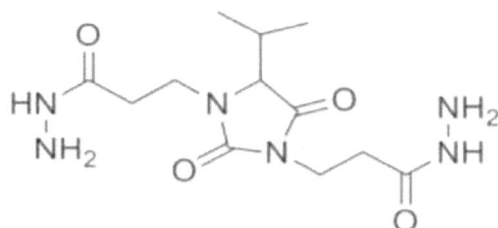

Crystalline powder
CAS. No.: 88122-32-1
Molar mass: 314
Melting point: 120°C

4-Isopropyl-2,5-dioxoimidazolidine-1,3-dipropionic acid dihydrazide (VDH)

Fig. (7). Chemical structure and physical properties of typical dihydrazide type curing agents.

Modified Imidazole Type Latent Curing Agents

Modified imidazole-type latent curing agents are supplied in fine powder with controlled particle size or as its premix in liquid epoxy resin [9-13]. Modified imidazole-type latent curing agents are typically prepared by grinding specially synthesized imidazole solids with a softening point ranging from 100 to 150°C. This modified imidazole incorporated epoxy resin composition is very stable at room temperature, with a pot life of up to 1 month because of significant lower-level contact opportunity between imidazole and epoxy group. Similar to the imidazole compound's cure mechanism as shown in Fig. (**1**), a modified imidazole type latent curing agent incorporated in thermal cure epoxy adhesives will dissolve and react with epoxide at elevated temperatures to form anionic species, which will initiate anionic polymerization of epoxy resin. The curing temperature can be very low.

Typical commercial-grade modified imidazole-type latent curing agent products are summarized in Table **3**. Ajicure PN series supplied by Ajinomoto Fine Techno. Co., Ltd. are mainly adduct products of imidazole and epoxy resin [14]. Ajicure PN-23 is the market standard product. Its composition with standard bisphenol A epoxy resin can cure fast at a relatively low temperature, around 80°C. The pot life is one month at 40°C. PN-23J is a finer particle size grade product of PN-23. PN-23J can cure faster than PN-23. The pot life is 2 weeks at 40°C. PN-H has better storage stability. Its pot life is 2 months at 40°C. Novacure HX series supplied by Asahi Kasei Corporation is a stabilized premix of imidazole adduct fine powder with liquid epoxy resin for easy use [15]. The typical usage level is 30 to 50 phr, higher than other powder-type products listed. HX-3722 is low-temperature curable grade. HX-3921HP has better storage stability with high-purity grade epoxy resin incorporated. Sunmide LH series supplied by Evonik Corporation are modified imidazole *via* Mannich reaction [16]. The composition of LH-210 with standard bisphenol A epoxy resin can cure fast at relatively low temperatures, around 80°C. The pot life is one month at 20°C. Cured epoxy resin possesses high glass transition temperature. LH-2102 is a faster cure version product with finer particle size. Adeka hardener EH series are adduct products of imidazole and epoxy resin incorporated with phenol resin [17]. EH-5011S is a low-temperature curable grade product. EH-5046S has a high glass transition temperature.

Table 3. Typical commercial-grade modified imidazole-type curing agents.

Product Name	Particle Size, μ	Softening Point, °C	Usage Level* Phr	Manufacturer
Ajicure PN-23	10-12	100-105	15-25	Ajinomoto Fine-Techno. Co., Inc.
Ajicure PN-23J	2-4	100-105	15-25	
Ajicure PN-H	11-13	120-125	15-25	

(Table 3) cont.....

Novacure HX-3722	2	-	30-50	Asahi Kasei Corp.
Novacure HX-3921HP	5	-	30-50	
Sunmide LH-210	5	130-140	10-25	Evonik Corp.
Sunmide LH-2102	2	130-140	10-25	
Adeka hardener EH-5011S	5	100-110	20-25	Adeka Corp.
Adeka hardener EH-5046S	5	120-125	20-25	

*Weight part based on 100 grams of standard bisphenol A epoxy resin with EEW around 190.

Typical cure behavior, physical properties, and adhesion performance of a simple bisphenol A epoxy resin formulated with Ajicure PN-23 are shown in Table **4** [18]. Gel time measured at 80°C was 13.2 minutes. The gel time measured on the hot plate was 16 seconds at 150°C and only 6 seconds at 180°C. This result indicates that modified imidazole formulated epoxy composition is possible for use in snap bonding required applications at suitable thermal cure conditions. Glass transition temperature measured by TMA was 139 °C. The epoxy composition showed good adhesion on steel.

Table 4. Thermal cure behavior and cured properties of imidazole adduct curing agent.

Formulation	
Standard Bisphenol A Epoxy Resin (EEW=190)	**100**
Ajicure PN-23	**25**
Aerosil 200	**1**
Gel time, min (2.5 g sample)	
@80°C	13.2
100°C	4.2

(Table 4) cont.....

120°C	2.2
Gel time on hot plate, seconds @150°C 180°C	16 6
Tg °C, TMA method Cured at 120°C x 60 min	139
Tensile strength, MPa Cured at 100°C x 60 min	46
Lap shear strength, MPa on steel Cured at 100°C x 60 min	12

Modified imidazole-type latent curing agents can also be used as an accelerator for other curing agents such as DICY, anhydrides, and phenols. The typical usage level is around 1 to 5 phr based on standard bisphenol A epoxy resin. Table **5** shows storage stability, cure behavior, physical properties, and adhesion strength of standard liquid bisphenol A epoxy resin incorporated with DICY and Ajicure PN-23 as compared to DCMU [19]. As can be seen, the modified imidazole-type curing agent shows a very good accelerating effect for DICY.

Table 5. Thermal cure behavior and cured properties of imidazole adduct as DICY accelerator.

Formulation				
Standard bisphenol A epoxy resin (EEW=190)	100	100	100	100
DICY	8	8	8	8
Ajicure PN-23	1	3	-	-
DCMU	-	-	3	-
	1	1	1	1

Aerosil 300				
Storage stability*, months @40°C	>12	>12	>12	>12
Gel time, min				
@120°C	43	14	24	uncured
150°C	6	2	4	uncured
Tg, °C, DMA method				
Cured at 120°C x 30 minutes	141	137	139	uncured
150°C x 30 minutes	147	132	140	uncured
150°C x 60 minutes	154	141	154	uncured
Lap shear strength, MPa on steel				
Cured at 120°C x 30 minutes	1	8	3	uncured
120°C x 60 minutes	5	18	28	uncured
150°C x 30 minutes	9	20	20	uncured
150°C x 60 minutes	28	24	24	uncured

*Storage period when viscosity becomes twice its initial value.

Modified Polyamine Type Latent Curing Agents

Modified polyamine-type latent curing agents are supplied in fine powder or premix in liquid epoxy resin. Modified polyamine-type latent curing agents are prepared by grinding specially synthesized modified aliphatic polyamine solid with a softening point ranging from 75 to 130°C. Modified polyamine-type latent curing agent incorporated with thermal cure epoxy adhesives dissolves and reacts with epoxide at elevated temperature *via* two routes: i) by forming anionic species, which will initiate anionic polymerization of epoxy resin; ii) by polymerizing *via* polyadditon reaction between active hydrogen and epoxy group. Its cure temperature is lowered significantly to 70°Cl. The formulated epoxy composition is very stable, with a pot life of typically longer than 1 month at room temperature.

Typical commercial-grade modified imidazole-type latent curing agent products are summarized in Table **6**. Adeka hardener EH-4357S and -5057PK supplied by Adeka Corporation are adduct products of aliphatic polyamine and epoxy resin incorporated with phenol resin [20]. They react with epoxy resin mainly *via* the polyaddition mechanism of active hydrogen in the modified polyamine and epoxy group, as illustrated in Fig. (**8**). Their composition with standard bisphenol A epoxy resin can cure fast at very low temperatures, around 70°C. The pot life is one month at 25°C. The usage level is around 50 to 60 phrs, mainly based on active hydrogen and epoxy equivalent calculation. Aradur 9506, supplied by Huntsman Corporation, is a polyamidoamine product of aliphatic polyamine and anhydride [21]. Aradur 940, supplied by Huntsman Corporation, is a premix of the fine particle modified polyamine curing agent and liquid epoxy resin for easy use. Their composition with standard bisphenol A epoxy resin can cure at low temperatures, around 100°C, with good storage stability at room temperature. They react with epoxy resin mainly *via* the polyaddition mechanism of active hydrogen in the modified polyamine and epoxy group. Usage level is also relatively high.

Table 6. Typical commercial-grade modified polyamine-type curing agents.

Product Name	Particle Size, μ	Softening Point, °C	Usage Level* Phr	Manufacturer
Adeka hardener EH-4357S	5	75-85	50-60	Adeka Corp.
Adeka hardener EH-5057PK	2	75-85	50-60	
Aradur 9506	-	105	30-50	Huntsman Corp.
Aradur 940	-	105	70-120	
Fujicure FXE-1020	4-6	120-130	20	T&K Toka Corp.
Fujicure FXE-1081	4-6	115-125	20	
Ajicure MY-24	8-10	115-120	15-25	

(Table 6) cont.....

Ajicure MY-H	8-10	125-130	15-25	Ajinomoto Fine-Techno. Co., Inc.

*Weight part based on 100 grams of standard bisphenol A epoxy resin with EEW of 190.

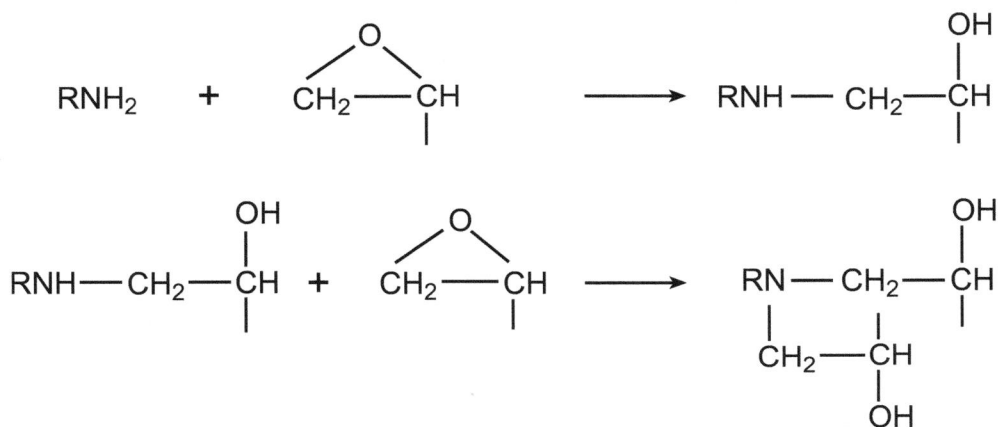

Fig. (8). Polyaddition polymerization of epoxy resin and modified polyamine.

Fujicure FXE-1020 and 1081 are mainly reaction products of aliphatic polyamine and urea [22]. They do not have enough active hydrogen remaining, so the main cure route is anionic polymerization of epoxy resin initiated by anionic species formed from the reaction of epoxy resin and tertiary amine of the modified polyamine latent curing agent dissolved at an elevated temperature. FXE-1020 with standard bisphenol A epoxy resin can cure at low temperatures, around 80°C. FXE-1081 with standard bisphenol A epoxy resin can cure at very low temperatures, around 70°C. The usage level is around 20 phr based on standard bisphenol A epoxy resin. Ajicure MY-24 and MY-H supplied by Ajinomoto Fine Techno. Co., Ltd. are adduct products of aliphatic polyamine and epoxy resin. Their cure route is also mainly anionic polymerization of epoxy resin initiated by anionic species formed from the reaction of epoxy resin and tertiary amine of the modified polyamine latent curing agent dissolved at elevated temperatures, as shown in Fig. (**9**). [23]. Their formulated epoxy resins can cure at temperatures around 100°C. They have very good storage stability. These four products can also be used as accelerators for curing agents such as DICY, anhydrates, and phenols.

Initiation

Propagation

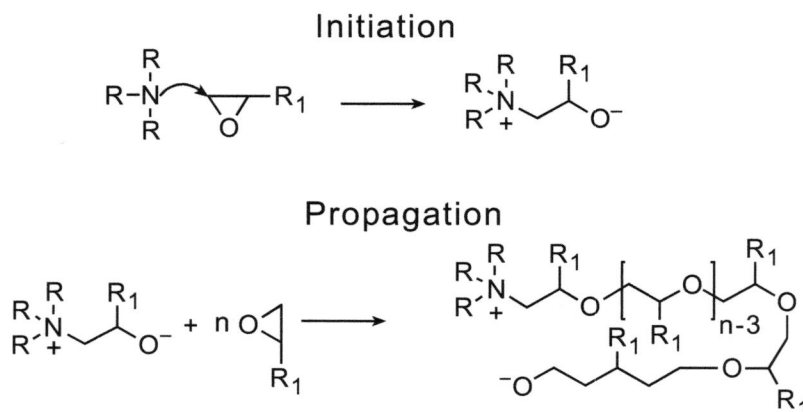

Fig. (9). Anionic polymerization of epoxy resin initiated by modified tertiary amine.

Thermal Cure Latent Cationic Initiators

Onium salt and BF_3-amine complex are the main commercialized thermolatent cationic initiators for epoxy resin. Typical commercial products are listed in Table 7.

Table 7. Typical commercial-grade thermal cure cationic initiators.

Product name	Chemical Type	Viscosity mPas/25°C	Usage Level* phr	Manufacturer
San Aid SI-60L	Onium salt solution	7	1.5 – 3	Sanshin Chemical Industrial Co., Ltd.
San Aid SI-100L		20	1 – 2	
K-Pure CXC-1612	Onium salt	Off-white powder	0.5 – 3	King Industries, Inc.
K-Pure CXC-1614		White crystal	0.5 – 3	
Anchor 1040	Amine-BF₃ complex	20000	5 - 10	Evonik Corp.
Anchor 1115		1700	7 - 12	

*Weight part based on 100 grams of standard bisphenol A epoxy resin with EEW of 190.

San Aid SI series supplied by Sanshin Chemical Co., Ltd. are aromatic sulphonium hexafluoro antimonate salt solutions (around 50% in high boiling point solvent) [24]. As illustrated in Fig. (**10**), [25-29] benzyl sulfonium salt initiator decomposes at elevated temperatures to generate benzylic cation species that can initiate cationic polymerization of epoxy resin. San Aid SI-60L-incorporated bisphenol A epoxy resin can cure at a relatively low temperature of around 90°C with good storage stability of up to 1 month pot life at room temperature. San Aid SI-100L-incorporated bisphenol A epoxy resin can cure at around 120°C with better storage stability. Cycloaliphatic epoxy resin incorporated compositions have much faster cure speed but shorter pot life because of the cationic cure mechanism. K-Pure CXC-1612 and CXC-1614, supplied by King Industries, Inc., are aromatic ammonium hexafluoro antimonate salt type initiators [30]. CXC-1612-incorporated epoxy resin can cure at a low temperature of around 80°C while CXC-1614-incorporated epoxy resin can cure at a temperature of around 100°C. Anchor 1040 and 1115, supplied by Evonik Corporation, are complex of amine and the boron trifluoride Lewis acid [31]. Cure temperature will be required to be higher, around 140°C.

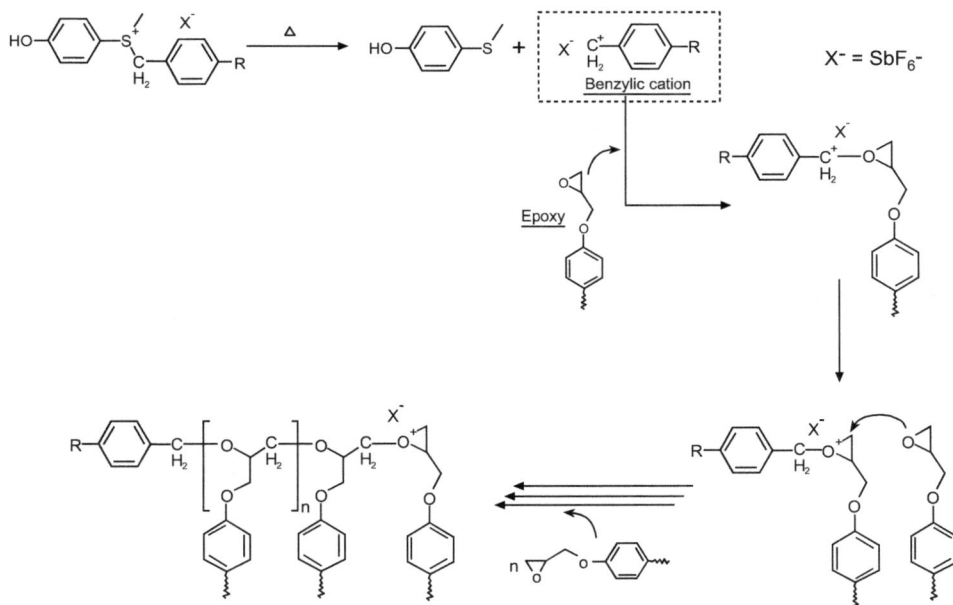

Fig. (10). Cationic polymerization of epoxy resin initiated by benzyl sulphonium salt type initiator.

ONE-COMPONENT THERMAL CURE EPOXY ADHESIVES

One-component thermal cure epoxy adhesives are primarily composed of epoxy resin, latent curing agent, and accelerator with various modifiers and additives. Key functions, ingredients, main role, and typical content are summarized in Table **8**.

Table 8. One-component thermal cure epoxy composition.

Function	Component	Main Role	Typical Content (%)
Primary	Epoxy resin	Adhesive base	20 – 95
	Latent curing agents	Curability	0.5 – 60
	Accelerator	Cure speed enhancement	0 – 5
Modifier	Filler	Property enhancement	0 – 80
	Toughener	Toughness enhancement	0 – 30
Additive	Colorant	Coloring	0 – 2
	Coupling agent	Adhesion promotion	0 – 2
	Thixotropic agent	Rheology control	0 – 5

Selection of a suitable type of latent curing agent is the key to formulating one-component thermal cure epoxy adhesives. The use of an accelerator or stabilizer might sometimes become necessary to achieve a higher balance of curability and stability. By the combination use of a liquid phenol compound, a special stabilizer with low-temperature curable latent curing agent, we found that the cure speed of one-component thermal cure epoxy adhesives can be significantly improved [32, 33]. As can be seen in Table **9**, the cure time at 99% conversion rate of sample no.2 was 6.75 minutes at 100°C as compared to 16.1 minutes for sample no. 4.

Table 9. Stability, cure behavior, and property comparison.

Sample No.	1	2	3	4	5
Composition content, %					
Bisphenol A epoxy resin[*1]	48.5	48.5	48.5	49.0	49.0
Latent curing agent[*2]	48.0	45.0	40.0	50.0	45.0
Liquid phenol resin[*3]	2.0	5.0	10.0	-	5.0
r-glycidoxy propyl trimethoxy silane	1.0	1.0	1.0	1.0	1.0
Stabilizer	0.5	0.5	0.5	-	-
Viscosity, mPa.s/25°C	24700	21900	25600	27200	25500
Pot life @25°C[*4], days	>14	>14	>14	>14	7
DSC cure time @100°C[*5], min @ 99% conversion rate	7.38	6.75	11.7	16.1	10.9
Modulus[*6], GPa@25°C	3.41	3.25	3.36	3.42	3.49
Tg, °C	154	151	146	159	152
Lap shear strength[*7], MPa	14.0	13.9	13.6	13.2	13.8

*1. RE-310S, manufactured by Nippon Kayaku Co., Ltd.

2. Novacure HX-3722, manufactured by Asahi Kasei Corp.

3. MEH 8000H, manufactured by Meiwa Plastic Industries, Ltd.

4. Pot life is measured by monitoring viscosity change at 25°Cistorage.

5. Conversion rate calculated according to DSC isothermal method.

6. Measured by DMA method. Sample cure condition: 100°C x 60 min.

7. Adherend substrate: glass-epoxy substrate. Sample cure condition: 100°C x 60 min.

Based on this study, a one-component thermal cure epoxy adhesive product, LOCTITE 3119, was developed and commercialized for general bonding applications. As can be seen from Fig. (**11**), this product has a very long pot life. There is completely no viscosity change when stored at 30°C for 2 weeks. This product cures rapidly at relatively low temperatures and provides good adhesion on a wide range of substrates. Typical properties are summarized in Table **10** [34].

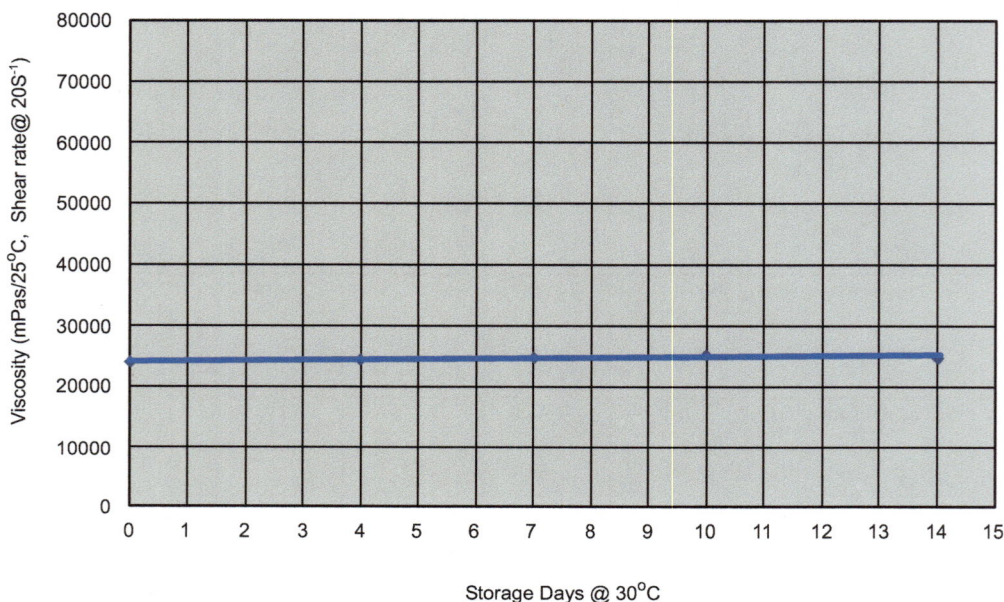

Fig. (11). Viscosity change of LOCTITE 3119 with storage condition at 30°C.

Table 10. Product description and typical properties of LOCTITE 3119.

Appearance	Black Viscous Liquid
Chemical type	One-component epoxy
Viscosity	14000 mPa.s/25°C
Density	1.2
Pot life	3 weeks @ 25°C
Hardness*, Shore D	87

(Table 10) cont.....

Tg, measured by DMA	110°C
Tensile strength	56 MPa
Tensile modulus	23 MPa
Elongation	3.3 %
Lap shear strength, on steel	24 MPa
Recommend cure condition	60 minutes @ 100°C 120 minutes @ 75°C
Storage condition	Frozen (-15 to -25°C)

*Cure condition: 100°C x 60 minutes.

A novel pure liquid one-component thermal cure epoxy adhesive was designed as a benchmark product to investigate the relationship between curability, workability, and stability of thermal cure epoxy composition [35, 36]. Its typical properties are shown in Table **11**.

Table 11. Typical properties of one-component pure liquid epoxy adhesive.

Appearance	**Light Yellow Liquid**
Chemical type	One-component pure liquid
Viscosity	6500 mPa.s/25°C
Tg, TMA method	121°C
Pot-life	15 hours @ 25°C
Cure condition	60 minutes @ 80°C 10 minutes @ 120°C
Storage condition	-40°C

Differential scanning calorimetry isothermal method is used to measure and analyze its cure behavior at 80°C, 90°C, 100°C, 110°C, 120°C, 130°C, 140°C, and 150°C. Fig. (**12**) shows the cure degree versus cure time at each temperature, calculated from DSC isothermal measurement results. Cure temperature and cure time at 90% cure degree are summarized in Table **12**.

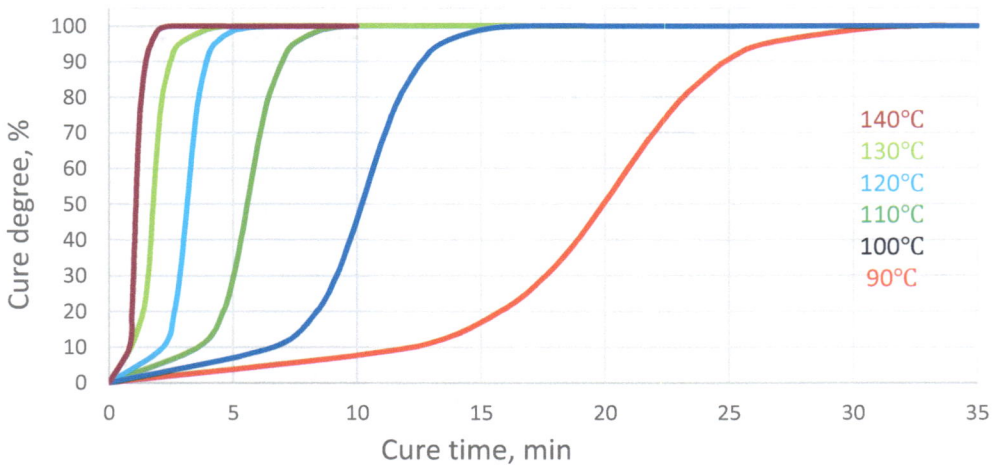

Fig. (12). Cure degree versus cure time, calculated from DSC Isothermal measurement results.

Table 12. Cure time at 90% cure degree versus cure temperature, measured by the DSC method.

Temperature, °C	1/Temp., K⁻¹	Cure Time, Min @ 90% Conversion	-Ln r, sec⁻¹
80	0.002831	49.8	8.00
90	0.002753	29.8	7.49
100	0.00268	12.6	6.62
110	0.00261	6.99	6.03
120	0.002543	3.9	5.46
130	0.00248	2.3	4.93

(Table 12) cont.....

| 140 | 0.00242 | 1.5 | 4.5 |
| 150 | 0.002363 | 1.0 | 4.09 |

Fig. (**13**) is the Arrhenius kinetic plot of -Ln r and 1/Temp, where r represents cure rate, calculated as 1/cure time in seconds. As can be seen, a very good straight line is obtained, indicating that the cure kinetic follows very well with Arrhenius theory. According to the kinetic equation shown below, a specific cure time of 90% cure degree can be calculated: 34 seconds @160°C and 14 seconds @180°C. In other words, the cure time to achieve a 90% conversion rate will decrease from 49.8 minutes @ 80°C to 3.9 minutes at 120°C, 60 seconds @150°C, and only 14 seconds at 180°C.

-Ln r = 8564.5 (1/Temp.) – 16.238

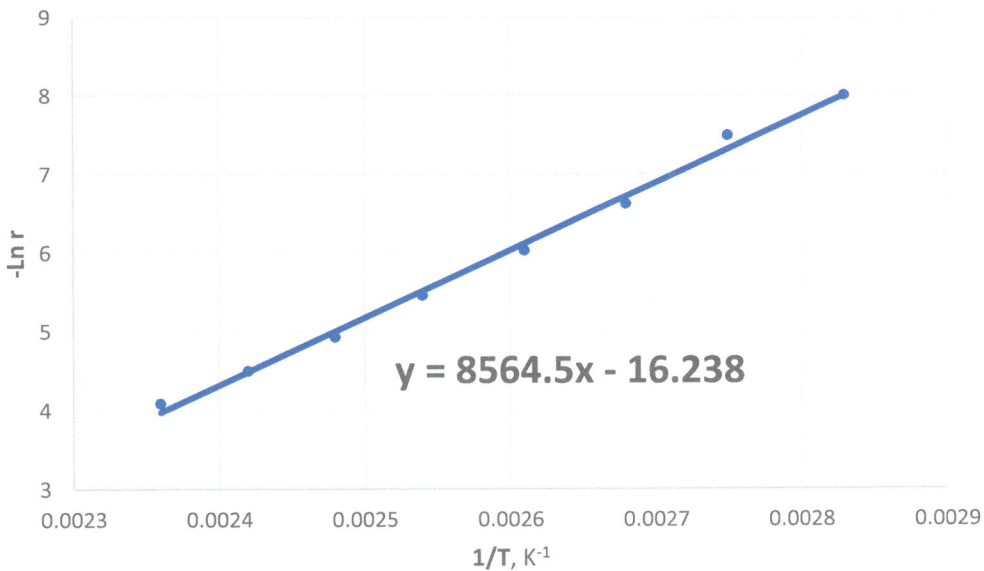

Fig. (13). Arrhenius kinetic plot.

Fig. (**14**) shows the viscosity change of this benchmark epoxy composition with storage of 25°C and -40°C. Based on cure behavior measurement results, the cure kinetics of the novel epoxy adhesive was investigated. Arrhenius equation obtained from kinetics plot curve was applied to predict pot life and shelf life. The result indicated that the predicted pot life and shelf-life correlate very well with actual measured results, as shown in Table **13**.

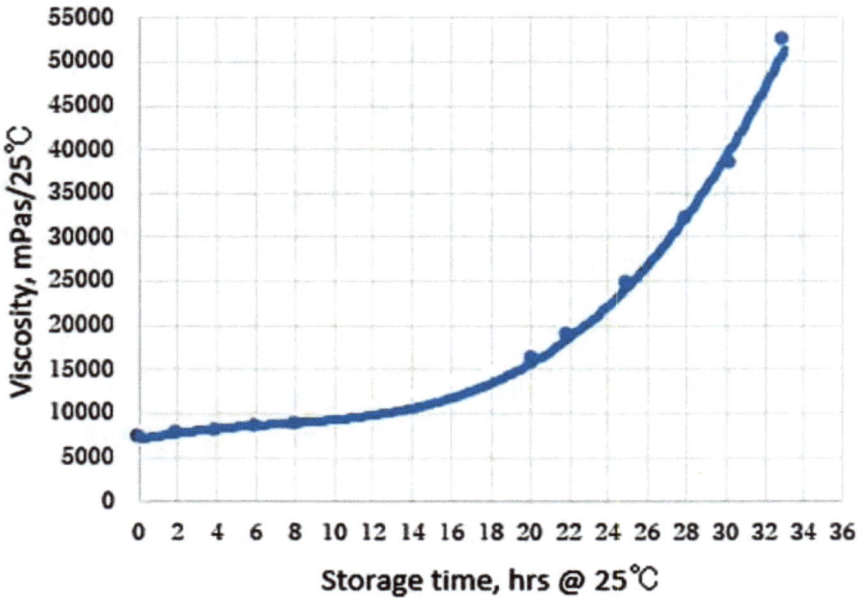

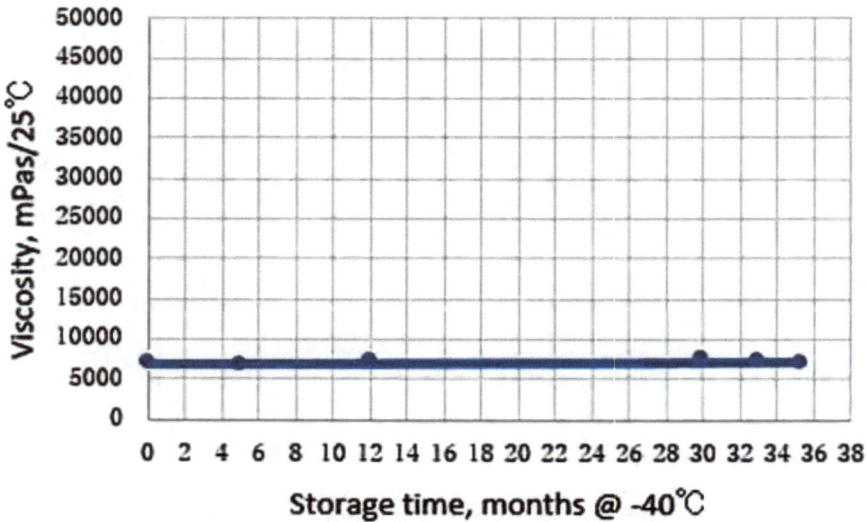

Fig. (14). Viscosity change, stored at 25°C (left), − 40°C (right).

Table 13. Calculated and measured shelf-life.

Storage condition	Shelf life, days	
	Calculated	Measured
@25°C	18.6 hours	19.2 hours
@5°C	5.4	5.6
@-20°C	93	82
@-40°C	1426	>1073

SNAP THERMAL CURE EPOXY BONDING TECHNOLOGY

Thermocompression bonding technology is a quick, easy method to reliably connect flip chips in microelectronics packaging and assembly applications. The flip chip bonding technology using adhesives has become more and more important because of its simple procedure, low-temperature process, and fine-pitch capability as compared to conventional processes using solder. There are two types of adhesives used: anisotropic conductive adhesives (ACAs) and non-conductive adhesives or non-conductive paste (NCA or NCP).

Anisotropic conductive adhesives include anisotropic conductive film (ACF) and anisotropic conductive paste (ACP) [37-48]. ACF technology is used in chip-on-glass (COG), flex-on-glass (FOG), flex-on-board (FOB), flex-on-flex (FOF), chip-on-flex (COF), and chip-on-board (COB) assemblies. ACP technology is mainly used in chip-on-flex (COF) applications. ACF used in COG bonding, FOG bonding, and FOB bonding for display module assembly is illustrated in Fig. (**15**) [49].

ACAs are typically one-component epoxy or acrylate composition. The epoxy-based ACF is primarily formulated from epoxy resin, latent curing agent, and fine conductive particles coated with a polymer. Table **14** shows one simple ACF composition example [50]. The ACF product is prepared by dissolving/dispersing the components in diethylene glycol mono-ethyl ether acetate and then screen printing the paste on PET film, followed by drying the film at 80°C for 5 minutes to evaporate the solvent. ACF bonding process usually has three steps: (1) Apply anisotropic conductive adhesives on the base substrate, (2) Mount the secondary substrate or device to the base substrate, and (3) Cure the adhesives at a certain high temperature with a press, usually at 170 to 200°C for 15 to 5 seconds with pressure applied at 0.5 to 4.0 MPa. ACF bonding structure and process are illustrated in Fig. (**16**). Typical ACF bonding parameters are summarized in Table **15**.

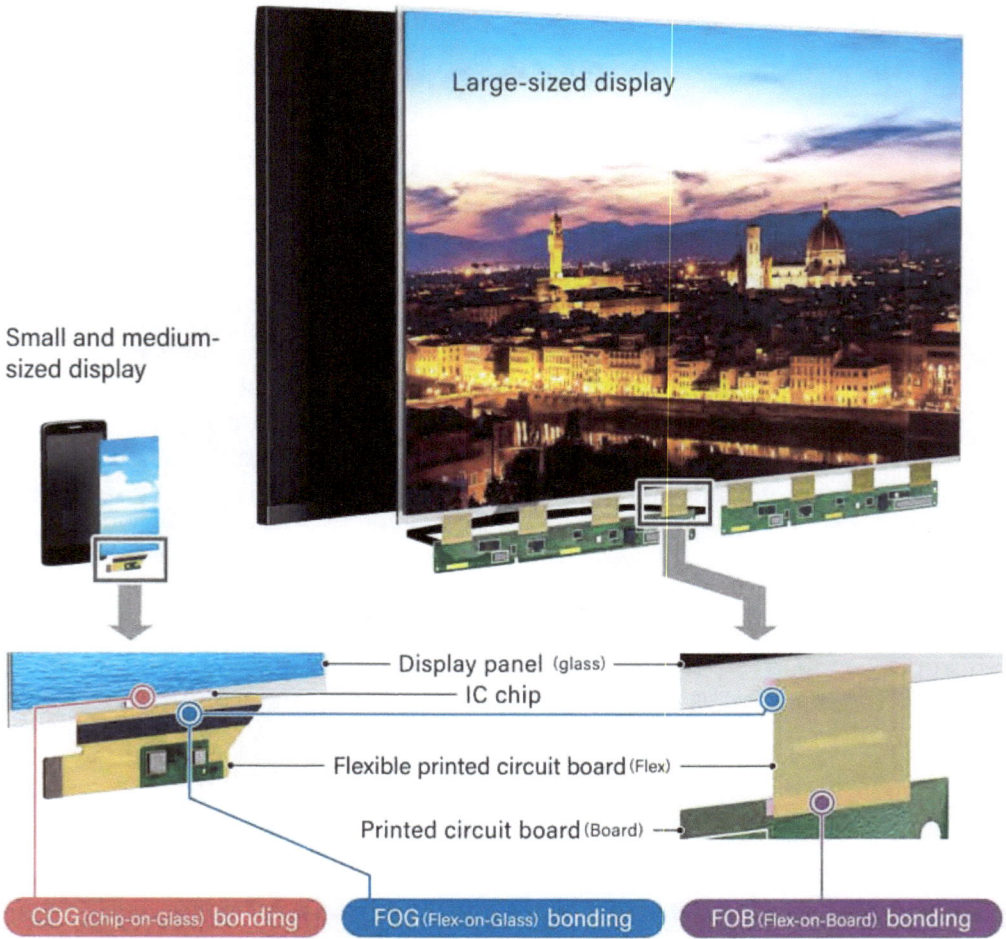

Fig. (15). ACF applications in display module assembly.

Table 14. ACF composition example.

Component	Weight %
Modified imidazole type latent curing agent premix with epoxy resin[1]	40
Phenoxy resin[2]	50
Polymer-coated conductive Au particles[3]	10

*1. Novacure HX 3941HP, manufactured by Asahi Kasei Corporation.

2. YP 50, manufactured by Toho Kasei Co., Ltd.

3. Micropearl AU, manufactured by Sekisui Chemical Co., Ltd.

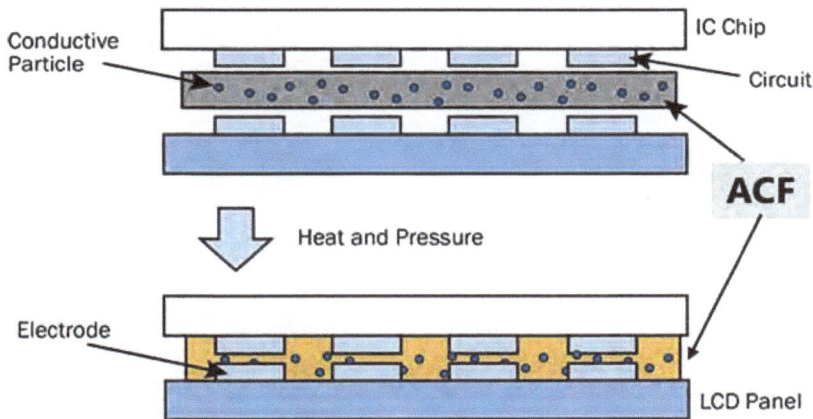

Fig. (16). ACF bonding structure and process.

Table 15. Typical ACF bonding parameters.

Assembly Type	Adhesive Base	Typical ACF Bonding Parameter		
		Pressure, MPa	Temperature, °C	Time, Seconds
FOG	Epoxy	2-4	170-200	10-12
COG	Epoxy	-	190-220	5-7
COF	Epoxy	-	190-220	5-10

(Table 15) cont.....

FOB	Epoxy	1-4	170-190	10-12
FOF	Acrylate	1-4	170-190	10-12
FOF	Acrylate	1-4	130-170	5-10

NCA or NCP is primarily composed of epoxy resin, latent curing agent, and inorganic fillers, with no conductive particle incorporated. NCA bonding technology is more suitable for use in finer pitch assembly [51-56]. Typical NCA bonding process is: (1) Apply nonconductive adhesive, (2) Mount the secondary substrate or device, (3) Temporarily cure the adhesives, usually at 170 to 240°C for 10 to 5 seconds with pressure applied at 0.5 to 4.0 MPa, and (4) Post thermal curing for full cure. NCA bonding structure and process are illustrated in Fig. (17).

Fig. (17). NCA bonding structure and process.

CONCLUSION

One-component thermal cure epoxy adhesives are typically formulated in the form of a compound containing epoxy resin, latent curing agent, and accelerator with a modifier and additive. New-type latent curing agents are modified imidazole or polyamines supplied in fine powder or as a premix in liquid epoxy resin. The selection of suitable latent curing agents is the key to formulating snap thermal cure epoxy adhesives. One-component thermal cure epoxy adhesives incorporated with these new type curing agents can cure very fast at certain elevated temperature conditions and have been successfully used as anisotropic conductive film (ACF), non-conductive paste (NCP) in semiconductor packaging, and electronics assembly applications.

REFERENCES

[1]　Chen, C. Structural epoxy adhesives In: *in: Structural Adhesives: Properties, Characterization and Applications*; Mittal, K.L. and Panigraphi, S.K., Eds.; Wiley – Scrivener: Beverly, MA, **2023**, pp. 3-30.
　　　http://dx.doi.org/10.1002/9781394175604.ch1

[2]　Ham, Y.R.; Kim, S.H.; Shin, Y.J.; Lee, D.H.; Yang, M.; Min, J.H.; Shin, J.S. A comparison of some imidazoles in the curing of epoxy resin. *J. Ind. Eng. Chem.,* **2010**, *16*(4), 556-559.
　　　http://dx.doi.org/10.1016/j.jiec.2010.03.022

[3]　Ikeda, Y.; Toyota, T. Basic curing agents and catalysts. In: *Epoxy resins*; Kakiuchi, H., Ed.; The Japan Epoxy Technology Society: Tokyo, **2003**, *Vol. 1,* pp. 147-155.

[4]　Odagiri, N.; Shirasu, K.; Kawagoe, Y.; Okabe, T. Cure path dependency in crosslinked structure of DGEBA/DICY/DCMU thermosetting resin. *Materials System,* **2021**, *38*, 21-30.

[5]　Tomuta, A.M.; Ramis, X.; Ferrando, F.; Serra, A. The use of dihydrazides as latent curing agents in diglycidyl ether of bisphenol A coatings. *Prog. Org. Coat.,* **2012**, *74*(1), 59-66.
　　　http://dx.doi.org/10.1016/j.porgcoat.2011.10.004

[6]　Lee, J.H. Using dihydrazides as thermal latent curing agents in epoxy-based sealing materials for liquid crystal displays. *Polymers (Basel),* **2020**, *13*(1), 109.
　　　http://dx.doi.org/10.3390/polym13010109 PMID: 33383914

[7]　Hirai, K. Latent curing agent. In: *High performance epoxy resin and curing agent formulation technology*; Kakiuchi, H., Ed.; Technology Information Association: Tokyo, **1997**, pp. 151-162.

[8]　Sato, N.; Inagaki, S.; Yamada, E. Structure and physical properties of epoxy resins reacted with various of polyurethanes. *J. Adhes. Sci. Technol.,* **2003**, *39*(8), 287-294.

[9]　Ajinomoto Fine-Techno Co., Ltd. Latent Curing Agent "AJICURE". Available at: https://www.aft-website.com/en/chemistry/ajicure [accessed on Dec 20, 2023].

[10]　Asahi Kasei Corporation. *NOVACURE*, Available at: https://www.asahi-kasei.co.jp/advance/en/business/ plastics/plastics04.html [accessed on Dec 20, 2023].

[11]　Evonik Corporation. *Epoxy additives and polyamides*, https://crosslinkers.evonik. com/product/ crosslinkers/downloads/Evonik-oxyguide asiapacific.pdf [accessed on Dec 20, 2023].

[12]　ADEKA Corporation. *ADEKA hardener EH series (latent hardener),* https://adeka-chemical.meclib.jp/adeka_epoxy_ENG/book [accessed on Dec 20, 2023].

[13]　Huntsman Corporation. *Advanced Materials*, https://www.huntsman.com/docs/ Documents/ US_2019_High_Performance_Components_Selector_Guide.pdf [accessed on Dec 20, 2023].

[14]　Takeuchi, K.; Abe, M.; Ito, N.; Hirai, K. *Latent curing agent for epoxy resin.* Japanese Patent Application S59-53526, **1984**.

[15]　Yamamoto, H.; Taiga, K. *Latent curing agent and its epoxy composition.* Japan Patent 4753934, **2011**.

[16] Tsuchiya, Y.; Hakuya, K.; Shiraishi, K. *Low temperature curable latent epoxy curing agent*. Japanese Patent 3390416, **2003**.

[17] Ogawa, R.; Yokoda, K.; Masamune, Y. *Latent curing agent and one-component epoxy composition*. Japan Patent 5876414, **2016**.

[18] Ohashi, J. Latent curing agent. In: *Epoxy resins*; Kakiuchi, H., Ed.; The Japan Epoxy Technology Society: Tokyo, **2003**, *Vol. 1*, pp. 186-197.

[19] Hirai, K. Latent curing agent. In: *High performance epoxy resins and formulation technology, evaluation and uses of curing agents*; Takaushi, K., Ed.; Technology Information Society: Tokyo, **1997**, p. 157.

[20] Fijita, Y. *Latent curing agent and one-component epoxy composition*. Japanese Patent 5248791, **2013**.

[21] Hirai, K. Latent curing agent. In: *High performance epoxy resins and formulation technology, evaluation and uses of curing agents*; Takaushi, K., Ed.; Technology Information Society: Tokyo, **1997**, p. 153.

[22] Otsuka, S. *Latent curing agent and one-component epoxy composition*. Japanese Patent 5377990, **2013**.

[23] Vidil, T.; Tournilhac, F.; Musso, S.; Robisson, A.; Leibler, L. Control of reactions and network structures of epoxy thermosets. *Prog. Polym. Sci.,* **2016**, *62*, 126-179.
http://dx.doi.org/10.1016/j.progpolymsci.2016.06.003

[24] Sanshin Chemical Industry Co., Ltd. *Cationic polymerization initiators,* https://www.sanshin-ci.co.jp/en/products.html [accessed on Dec 20, 2023].

[25] Koike, T. Structure and properties of thermal cationic polymerization initiator as latent curing agent for epoxy resin. *Journal of The Adhesion Society of Japan,* **2020**, *56*(1), 20-33.
http://dx.doi.org/10.11618/adhesion.56.20

[26] Shimomura, O.; Tomita, I.; Endo, T. Curing behavior of epoxy resin initiated byS-alkylsulfonium salts of aromatic sulfides as thermal latent cationic initiators. *J. Polym. Sci. A Polym. Chem.,* **2001**, *39*(6), 868-871.
http://dx.doi.org/10.1002/1099-0518(20010315)39:6<868::AID-POLA1060>3.0.CO;2-4

[27] Shimomura, O.; Tomita, I.; Endo, T. Application of novel polymers with *S* -alkylsulfonium salt moieties as alkylating agents and thermal latent cationic initiators. *J. Polym. Sci. A Polym. Chem.,* **2001**, *39*(22), 3928-3933.
http://dx.doi.org/10.1002/pola.10036

[28] Nakano, S.; Endo, T. Thermal cationic curing with benzylammonium salts — 2. *Prog. Org. Coat.,* **1996**, *28*(2), 143-148.
http://dx.doi.org/10.1016/0300-9440(95)00612-5

[29] Park, S.J.; Seo, M.K.; Lee, J.R.; Lee, D.R. Studies on epoxy resins cured by cationic latent thermal catalysts: The effect of the catalysts on the thermal, rheological, and mechanical properties. *J. Polym. Sci. A Polym. Chem.,* **2001**, *39*(1), 187-195.
http://dx.doi.org/10.1002/1099-0518(20010101)39:1<187::AID-POLA210>3.0.CO;2-H

[30] King Industries, Inc. *K-Pure CXC-1612*, https://www.kingindustries.com/assets/1/6/K-PURE_CXC-1612_PDS [accessed on Dec 20, 2023].

[31] Evonik Corporation. *Anchor 1040*, https://www.productfinder.crosslinkers.com/pdf/daten/Anchor_1040_e.pdf [accessed on Dec 20, 2023].

[32] Chen, C. Thermal curable liquid resin composition. *Japanese Patent 4204814*, **2002**.

[33] Chen, C. One-component epoxy resin composition. *European Patent 2640765*, **2017**.

[34] Evonik Corporation. *Loctite Ablestik 3119*, https://www.henkel-adhesives.com/jp/ja/product/electrically-non-conductive-adhesives/loctite_ablestik31190.html [accessed on Dec 20, 2023].

[35] Chen, C.; Li, B.; Kanari, M.; Lu, D. Curability, workability and stability investigation on a novel pure liquid one component thermal curable epoxy adhesive. *Adv. Mater.*, **2019**, *8*(2), 94-99.
http://dx.doi.org/10.11648/j.am.20190802.16

[36] Chen, C.; Li, B.; Wang, C.; Iwasaki, S.; Kanari, M.; Lu, D. *Pure liquid one part thermal cure epoxy adhesive.* **2018**.

[37] Shiiki, M.; Imaizumi, J.; Miyata, T.; Chinda, A. Materials and components for flat panel display applications. *Hitachi Rev.*, **2006**, *55*, 32-39.

[38] Toshioka, H.; Nakatsugi, K.; Yamamoto, M.; Sato, K.; Shimbara, N.; Okuda, Y. Development of anisotropic conductive film for narrow pitch circuits. *SEI Technical Rev.*, **2011**, *73*, 41-44.

[39] Takano, N.; Fujinawa, T.; Kato, T. Film technologies for semiconductors and electronic components. *Hitachi Chemical Technical Report*, **2013**, *55*, 20-23.

[40] Licari, J.J.; Swanson, D.W. *Adhesives Technology for Electronic Applications: Material, Processing, Reliability*, 2nd ed; Elsevier: New York, **2011**, pp. 6-8.

[41] De Vries, J.W.C.; Caers, J.F.J.M. Anisotropic conductive adhesives in electronics. In: *Advanced Adhesives in Electronics*; Alam, M.O.; Bailey, C., Eds.; Woodhead Publishing: Cambridge, UK, **2011**, pp. 53-104.
http://dx.doi.org/10.1533/9780857092892.1.53

[42] Tao, B.; Ding, H.; Yin, Z.; Xiong, Y. ACF curing process optimization for chip-on-glass (COG) considering mechanical and electrical properties of joints. In: *New Developments in Liquid Crystal*; Thachenko, G.V., Ed.; IntechOpen: London, UK, **2009**, pp. 179-206.
http://dx.doi.org/10.5772/9688

[43] Kim, J.Y.; Kim, E.R.; Ihm, D. Anisotropic conductive film (ACF) prepared from epoxy/rubber resins and its fabrication and reliability for LCD. *J. Inf. Disp.*, **2003**, *4*(1), 17-23.
http://dx.doi.org/10.1080/15980316.2003.9651908

[44] Sakairi, M. Bonding materials. *U.S. Patent 6576334*, **2003**.

[45] Akutsu, Y. Anisotropic electrically conductive film. *U.S. Patent 10026709*, **2018**.

[46] Tatsuzawa, T. Anisotropic conductive film and method for producing the same. *Japanese Patent 4650490*, **2010**.

[47] Yoon, D.J.; Malik, M.H.; Yan, P.; Paik, K.W.; Roshanghias, A. ACF bonding technology for paper- and PET-based disposable flexible hybrid electronics. *J. Mater. Sci. Mater. Electron.*, **2021**, *32*(2), 2283-2292.
http://dx.doi.org/10.1007/s10854-020-04992-2

[48] Lin, C.M. Electrical resistance effects of anisotropic conductive film-assembled flex-on-flex packages under static bending loads. *Microsyst. Technol.*, **2018**, *24*(6), 2577-2584.
http://dx.doi.org/10.1007/s00542-017-3663-7

[49] Dexerials Corporation. *The fundamentals of anisotropic conductive film (ACF)*, https://techtimes.dexerials.jp/en/bonding/what-is-acf/ [accessed on Dec 20, 2023].

[50] Takeuchi, M.; Yamada, Y.; Suga, Y. Anisotropic conductive adhesive. *Japanese Patent 3982444*, **2007**.

[51] Shimote, Y.; Iwasaki, T.; Watanabe, M.; Baba, S.; Kimura, M. The fine pitch Cu-pillar bump interconnect technology utilizing NCP resin, achieving the high quality and reliability. J. Japan Inst. *Electron. Packag.*, **2014**, *7*, 87-93.

[52] Gim, M.; Kim, C.; Na, S.; Ryu, D.; Park, K.; Kim, J. High-performance flip chip bonding mechanism study with laser assisted bonding *Proceedings of 2020 IEEE Electronic Components and Technology Conference,* **2020**, pp. 1025-1030.
http://dx.doi.org/10.1109/ECTC32862.2020.00166

[53] Mackie, A.; Jo, H.; Lim, S.P. Flip-chip flux evolution, in: *Proceedings of IMAPS 2019 – 52nd International Symposium on Microelectronics*, **2019**, 115-119.
http://dx.doi.org/10.4071/2380-4505-2019.1.000115

[54] Kim, S.C.; Kim, Y.H. Review paper: Flip chip bonding with anisotropic conductive film (ACF) and nonconductive adhesive (NCA). *Curr. Appl. Phys.*, **2013**, *13*, S14-S25.
http://dx.doi.org/10.1016/j.cap.2013.05.009

[55] Chung, C.; Paik, K. Non-conductive films (NCFs) with multi-functional epoxies and silica fillers for reliable NCFs flip chip on organic boards (FCOB *Proceedings of 2007 Electronics Components and Technology Conference,* **2007**, pp. 1831-1838.
http://dx.doi.org/10.1109/ECTC.2007.374046

[56] Saarinen, K.; Frisk, L. Changes in adhesion of non-conductive adhesive attachments during humidity test. IEEE Tran. *IEEE Trans. Compon. Packaging Manuf. Technol.,* **2011**, *1*(7), 1082-1088.
http://dx.doi.org/10.1109/TCPMT.2011.2141992

CHAPTER 5

Induction Cure Epoxy Technology

Abstract: Induction cure epoxy technology is based on the induction heating method. Induction heating is a very fast, highly efficient, and noncontact heating style. Induction heating works mainly on bonding applications on conductive metal materials or using specially designed induction curable epoxy adhesives. The induction cure principle and equipment, induction cure epoxy chemistry, and induction cure applications are described. In addition, laser cure epoxy adhesive technology and weld bonding epoxy adhesive technology have been briefly introduced.

Keywords: Conductive metal, Induction cure, Induction coil, Induction heating, Thermal conductivity.

INDUCTION CURE EPOXY CHEMISTRY

Induction cure epoxy technology is a technology based on the induction heating method to cure epoxy materials. Induction heating is a form of electromagnet heating. It is a very fast, highly efficient, and non-contact heating. Electric power is used to generate heat *via* conductive metal materials placed in an inductor coil of induction cure equipment. Induction heating works only on bonding applications where one substrate is a conductive metal material, such as steel, aluminum, or copper, or specially designed induction curable adhesives filled with conductive metal powders. The advantages of induction cure epoxy technology are listed below:

✓Instant curability because of very fast heating speed.

✓Thermal-sensitive material bonding because of localized heating.

✓High adhesion performance.

As shown in Fig. (**1**), there are two main cases in which epoxy adhesives are used to bond the substrates, depending on the adherend type [1]. If one of the adherends is metal that can generate induction heating, the heat generated will conduct thermally to the epoxy adhesive that will start to cure soon after a certain temperature is reached. The heating rates could be very fast, normally higher than 40°C/second, and thus can reach 190 – 230°C in a few seconds, where most epoxy adhesives can cure instantly. Metals that have been used in induction heating are iron, steel, copper, and aluminum. Typically, one-component epoxy adhesive is

Chunfu Chen

used for induction cure applications because of easy handling, while there are some applications using two-component epoxy adhesive.

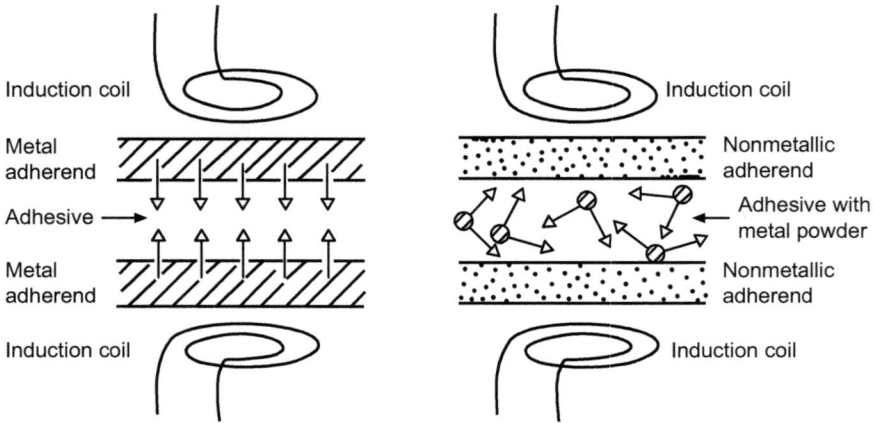

Fig. (1). Induction cure epoxy applications. Left: normal epoxy adhesive used in bonding metal adherend; Right: epoxy adhesive filled with metal powder in bonding non-metallic adherend.

In non-metallic adherend applications, specially designed induction curable epoxy adhesive, which is filled with enough amount of susceptor that can generate induction heating, needs to be used. Table **1** shows the typical composition of induction curable epoxy adhesive. Induction cure epoxy adhesive is fundamentally composed of epoxy resin, curing agent, and susceptor. Susceptors that can be used to generate induction heating are iron powder, steel powder, copper powder, and aluminum powder [2-10]. Aluminum powder is most commonly used due to its relatively low density. Fig. (**2**) shows a typical aluminum powder product and its micrograph. Overheating is always a big concern in induction cure applications. Precautions need to be made to avoid high cure exotherm in designing the adhesive formulation as the cure exotherm will significantly increase the heating temperature.

Table 1. Typical composition of induction cure epoxy adhesive.

Function	Component	Main Role	Typical Content (%)
Primary	Epoxy resin	Adhesive base	20 – 95
	Curing agents	Curability	0.5 – 60
	Accelerator	Cure speed enhancement	0 – 5

Susceptor	Metal powder	Induction heating generation	10-80
Modifier	Filler	Property enhancement	0 – 80
	Toughener	Toughness enhancement	0 – 20
Additive	Colorant	Coloring	0 – 2
	Coupling agent	Adhesion promotion	0 – 2
	Thixotropic agent	Rheology control	0 – 5

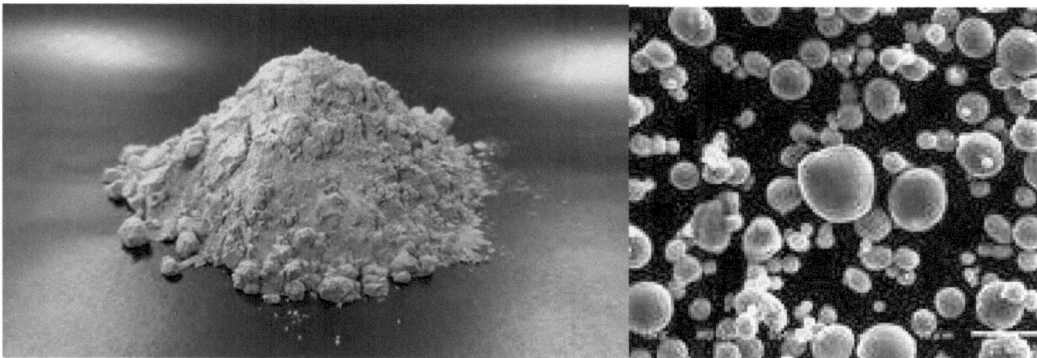

Fig. (2). Aluminum powder product.

INDUCTION CURE EQUIPMENT

Physical Principle of Induction Heating

The phenomenon of electromagnetic induction heating is based on three physical principles: 1) Transfer of energy from the inductor *via* electromagnetic fields, 2) Transformation of the electric energy to heat, and 3) Transmission of the heat inside the object *via* thermal conduction.

1) Transfer of energy from the inductor to the object by means of electromagnetic fields.

The electromagnetic field generated *via* a solenoid coil is illustrated in Fig. (3) [11]. The magnetic flux density B can be applied by Ampere Circuital Law, as shown below. Here, N is the number of turns, ℓ is the solenoid length, μ_0 is the magnetic constant, and I is the current.

$$B = \mu_0 \frac{N}{l} I$$

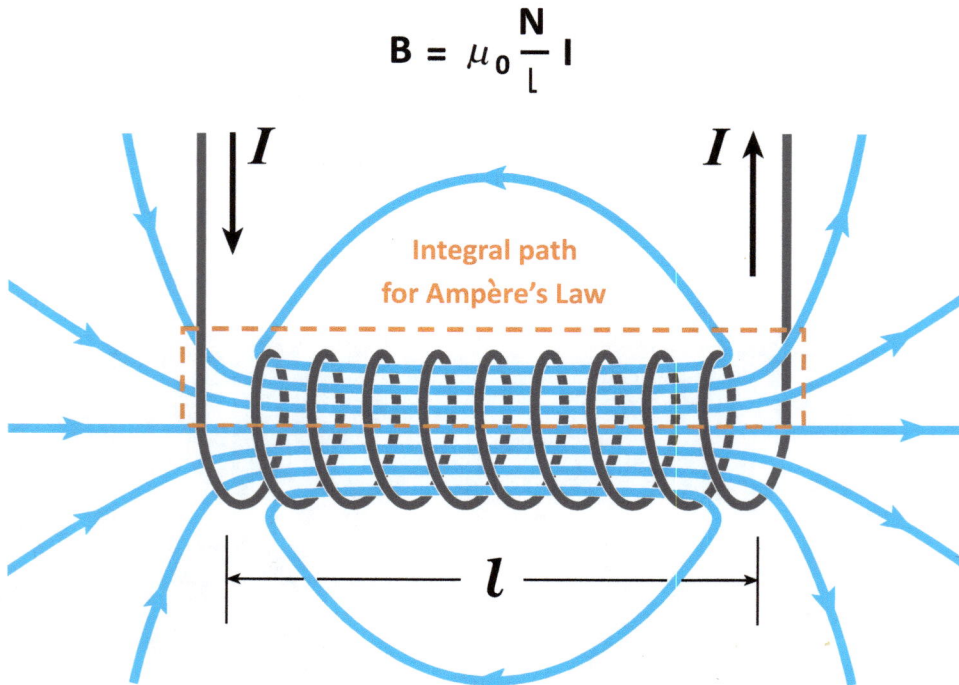

Fig. (3). Electromagnetic field generation of a solenoid coil.

2) Transformation of the electric energy into heat due to the Joule effect.

Joule heating is the physical effect by which the passing of current through an electrical conductor produces thermal energy that will result in a rise in the conductor material temperature. Thus, Joule heating is a transformation between "electrical energy" and "thermal energy", following the energy conservation principle theoretically shown in the below equation:

$$\frac{H}{t} = I^2 R$$

where:

- H is the heat produced by the conductor, in Joules;
- I is the electrical current flowing through the conductor, in amperes;
- R is the electrical resistance, in ohms;
- t is the elapsed time, in seconds.

Skin effect is the tendency of an alternating electric current (AC) to become distributed within a conductor such that the current density is largest near the surface of the conductor and decreases exponentially with greater depths in the conductor. It is caused by opposing eddy currents induced by the changing magnetic field resulting from the alternating current. The skin effect decreases with an increase in frequency, as illustrated in Figs. (**4, 5**) shows skin depth *vs.* frequency for some metal materials at room temperature. The red vertical line denotes 50 Hz frequency, where Mn-Zn is magnetically soft ferrite, Al is metallic aluminum, Cu is metallic copper, steel 410 is magnetic stainless steel, Fe-Si is grain-oriented electrical steel, and Fe-Ni is high-permeability permalloy (80%Ni-20%Fe).

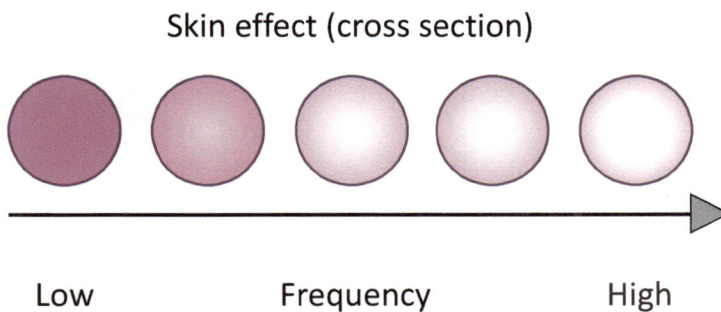

Fig. (4). Skin effect *versus* frequency.

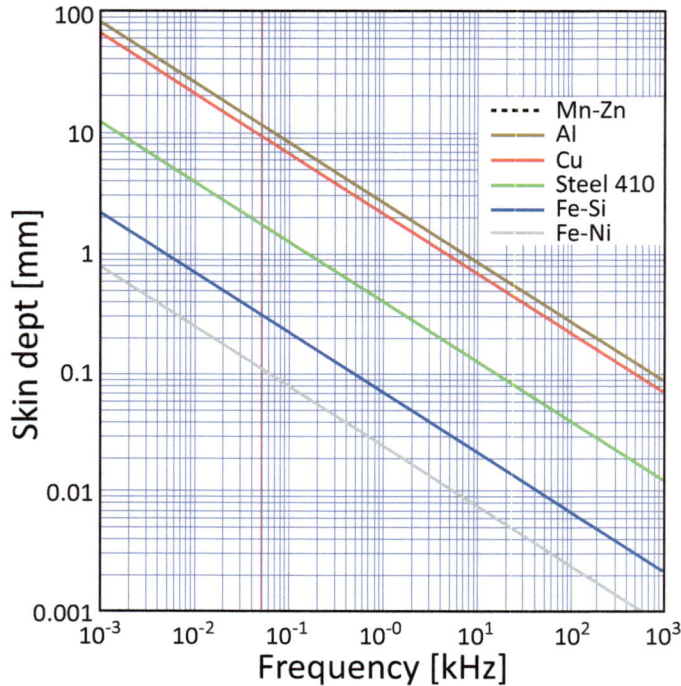

Fig. (5). Skin depth *vs.* frequency for some materials at room temperature.

Induction heating is generated according to eddy current loss and/or hysteresis loss of conductive metals. Eddy current loss, also known as Foucault or joule loss, is a phenomenon that occurs when a conductor is exposed to a varying magnetic field, resulting in the generation of circulating currents within the conductor. These circulating currents are known as eddy currents and are responsible for the dissipation of energy in the form of heat. Hysteresis loss, also known as magnetic hysteresis loss, is a phenomenon that occurs in ferromagnetic materials when they are subjected to cyclic magnetic fields. It refers to the energy dissipated in the form of heat as the magnetic domains within the material undergo repeated alignment and realignment with the changing magnetic field. Typical metals that can be used as induction heating material are listed in Table **2**. Common induction heating metals are copper, brass, aluminum, iron, and stainless steel [12]. Gold and silver are very good conductive materials but are not suitable for induction heating due to their low resistivity.

Table 2. Typical induction heating metal materials.

Material	Resistivity, Ω·m	Magnetic Permeability, H/m	Eddy Current Loss	Hysteresis Loss
Copper	1.7	1	+++	NG
Brass	5-7	1	+++	NG
Aluminum	2.8	1	+++	NG
Lead	21	1	+++++	NG
Stainless steel (nonmagnetic)	10-50	1	+++++	NG
Stainless (magnetic)	10-50	50-100	+++++	+++++
Iron	10-20	100-200	+++++	+++++

3) Transmission of the heat inside the object by means of thermal conduction.

Thermal conduction of materials is applied by Fourier's Law as shown in the below equation. The heat flux density q is equal to the product of thermal conductivity k and the negative local temperature gradient $-\nabla T$. Table **5** lists thermal conductivity data of various types of epoxy adhesives, inorganic fillers, and metals.

$$\mathbf{q} = -k\nabla T,$$

where (including the SI units)

- $\mathbf{q}$ is the local heat flux density, W/m^2,
- k is the material's conductivity, W/(m·K),
- ∇T is the temperature gradient, K/m.

Table **3** shows the chemical composition, density, and coefficient of thermal conductivity of various polymers, inorganic fillers, and metals. Table **4** shows the thermal conductivity of different types of epoxy adhesives [13-16]. Fig. (6) illustrates the coefficient of thermal conductivity of various materials [17].

Table 3. Coefficient of thermal conductivity of polymers, inorganic fillers, and metals.

Filler	Chemical Composition	Density	Coefficient of Thermal Conductivity, W/m.K
Polypropylene (PP)	-	0.89 – 0.92	0.18-0.24
Polystyrene (PS)	-	0.96 – 1.05	0.13
Polyvinyl chloride (PVC)	-	1.3 – 1.45	0.21
Nylon	-	1.14	0.30
Epoxy resin	-	-	0.17-0.21
Thermally conductive epoxy adhesive	-	-	1 - 5
Calcium carbonate	$CaCO_3$	2.71	2.3
Silica	SiO_2	2.65	2.9
Kaolin clay	$Al_2Si_2O_5(OH)_4$	2.65	2.0
Talc	$Mg_3Si_4O_{10}(OH)_2$	2.7	2.1
Silica carbide	SiC	3.21	270
Alumina	Al_2O_3	3.95	30
Stainless steel	-	7.8	16
Carbon steel	-	7.8	36
Brass	-	8.4 – 8.73	109
Aluminum	Al	2.7	234
Copper	Cu	8.96	386
Silver	Ag	10.49	427

(Table 3) cont.....

Graphite	C	2.26	100-400
Carbon nanotube	C	2.26	1000-4000
Diamond	C	3.5	1000~2000

Table 4. Thermal conductivity of commercial epoxy adhesives.

Material	Filler	Product Name	Thermal Conductivity, W/m.K
Epoxy paste adhesive	Silver	Ablebond 84-1 LMI SR4[1]	2.5
	Silver	Abletherm 2600BT[1]	20
	Silver	ESP 8680-WL[2]	8
	Silver	ME8455-ML2[2]	6.5
	Silver	KO 110[1]	3.5
	Gold	Epo-Tek H43[3]	1.7-2
	Gold	Ablebond 85-1[1]	2
	Copper	Tru Bond 215[4]	0.98
	Diamond	ME 7156[2]	11.5
	Alumina	Ablebond 84-3 MVB[1]	0.5
Low stress epoxy paste	Alumina	MEE 7655[2]	1.8
	Alumina	ME-7155[2]	1.7
Epoxy adhesive	silica	LCA-BA5[5]	0.83
	Boron nitride	Epo-Tek 930[3]	4.1
	Glass fabric	Ablefilm 504[1]	0.78

(Table 4) cont.....

Epoxy film, thermally conductive	Glass fabric	Ablefilm 5020K[1]	0.7

*1. Supplied by Henkel Corporation.

2. Supplied by AI Technology Incorporation.

3. Supplied by Epoxy Technology, Incorporation.

4. Supplied by ITW Performance Polymers.

5. Supplied by Bacon Industries.

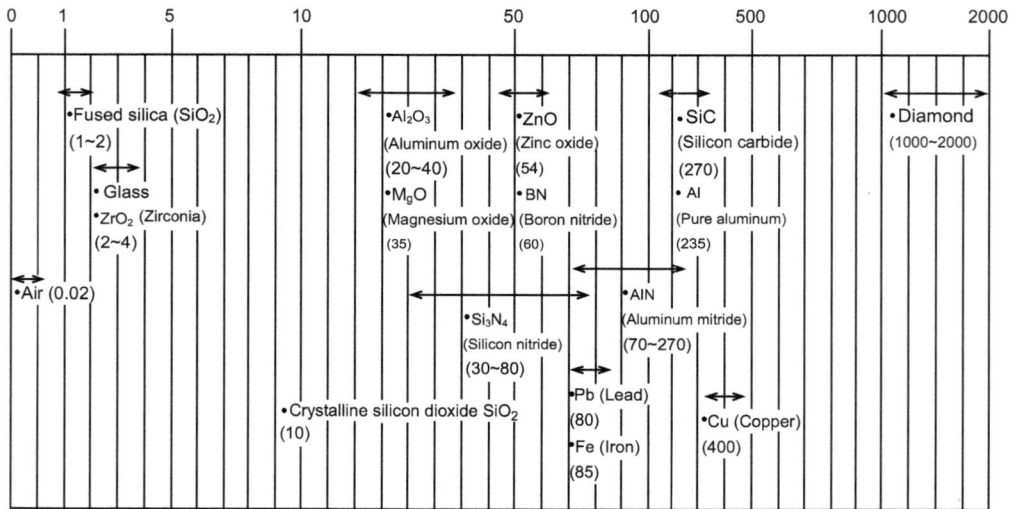

Fig. (6). Coefficient of thermal conductivity chart of various materials in W/m.K.

Induction Cure Equipment

As shown in Fig. (**7**), induction cure equipment is mainly composed of three units: generator, matching unit, and work coil. A generator is a power unit that can generate and convert to a useful range of electric power and frequency. A matching unit is a combination of a capacitor and transformer that is used to match the power unit and work coil. A work coil is used to heat the object/substrate. Fig. (**8**) illustrates a coil design with copper tubes for a cooling effect. Fig. (**9**) shows actual induction cure equipment supplied by Seidensha Electronics Co., Ltd [18].

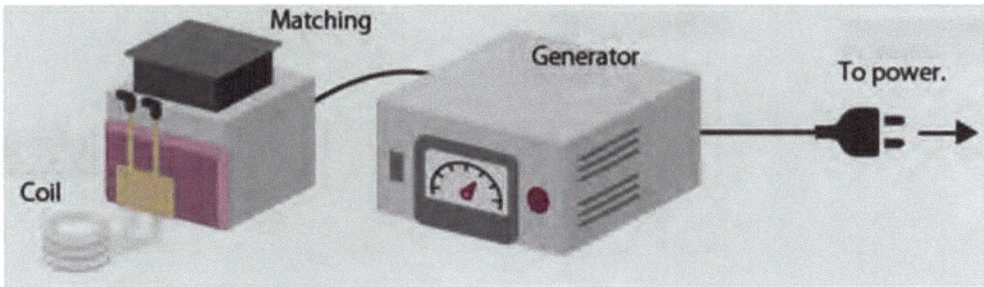

Fig. (7). Typical induction cure equipment structure.

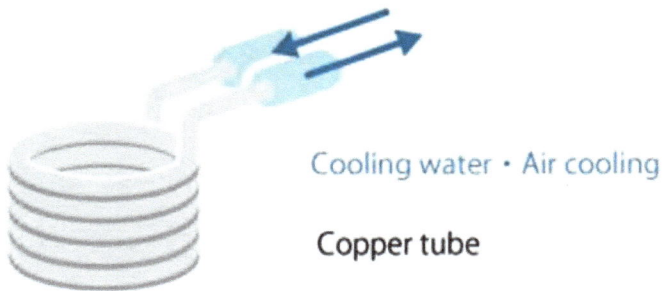

Cooling water · Air cooling

Copper tube

Fig. (8). Work coil design.

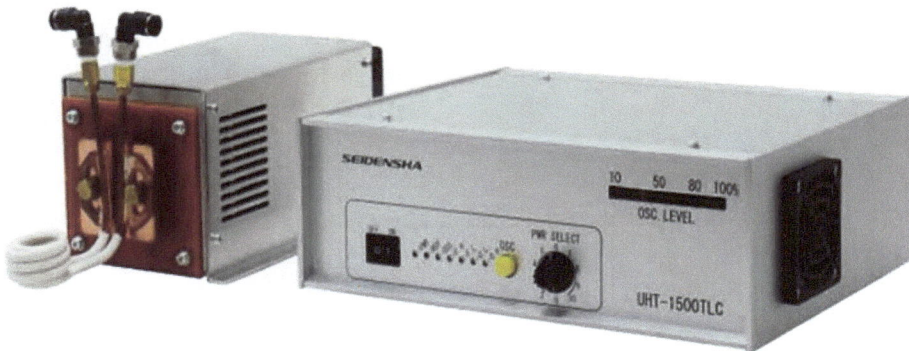

Fig. (9). Induction cure equipment.

Induction heating is very efficient, and the temperature increase can be very fast. Fig. (**10**) shows an example of temperature-raising speed of induction heating in comparison with standard oven cure. As can be seen clearly, the heating time of

180°C was shorted to a few seconds in induction curing and dozens of minutes in oven curing.

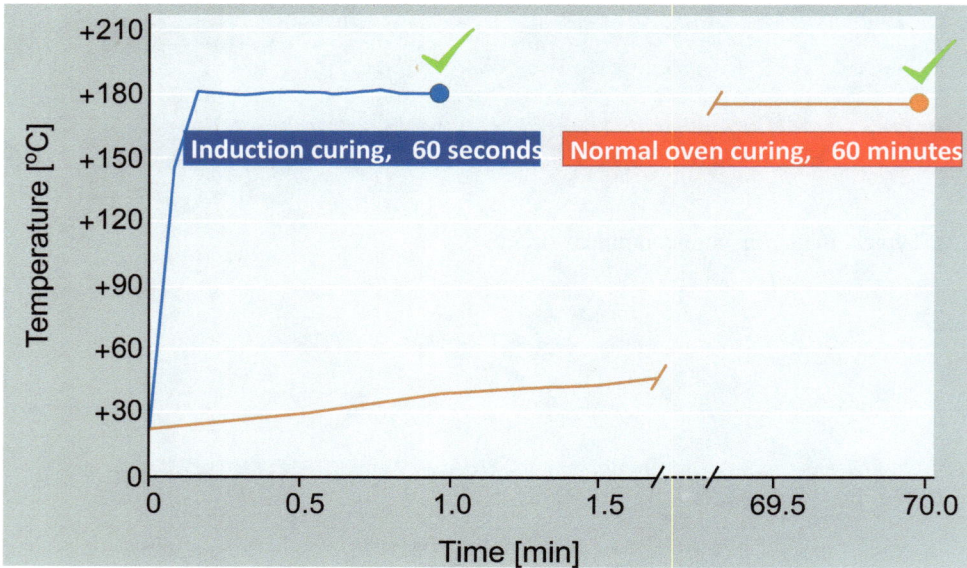

Fig. (10). Induction heating vs standard oven heating.

APPLICATIONS

Induction cure epoxy adhesives have been commercialized by several epoxy adhesive suppliers [19-22]. Typical applications are magnet bonding for e-motor assembly, hem flange bonding in the automotive industry, and general-purpose bonding applications that require fast production efficiency. Table **5** shows the typical properties of a one-component induction cure epoxy adhesive Loctite EA E-220IC, supplied by Henkel Corporation [23]. Table **6** compares the results of its lap shear adhesion strength, cured by induction cure condition and normal oven cure condition. As can be seen, the product shows very satisfactory adhesion performance under induction cure conditions.

Table 5. Typical properties of LOCTITE EA E-220IC.

Appearance	Grey Paste
Viscosity, mPa.s/25°C	90500
Glass transition temperature* (Tg), °C	87

(Table 5) cont.....

Shore hardness*	>80
Volume shrinkage*, %	0.5
Elongation* at break, %	3
Tensile strength* at break, MPa	48
Dielectric breakdown strength*, kV/mm	2.3
Cure	Heat cure or induction cure
Application	Bonding

*Cured at 125°C x 60 min.

Table 6. Adhesion performance comparison.

Lap Shear Strength, MPa	Induction Cure Cured at 90 amps x 90 sec., Peak Temperature: 230°C	Oven cure Cured at 125°C x 60 min
Steel (grit blasted)	20	41
Stainless steel	22	33
Aluminum (abraded)	23	24
Aluminum (anodized)	12	24
Nylon to steel (grit blasted)	1.5	2
Wood to steel (grit blasted)	5	4
Block shear strength, MPa Ferrite magnet to steel	23	13

Induction cure epoxy bonding technology has become an important method for curing adhesives and sealants for car components assembly in the automotive industry. Fig. (**11**) shows a typical hem flange bonding process [24]. Epoxy adhesive is applied before joint forming and is partially cured *via* induction heating.

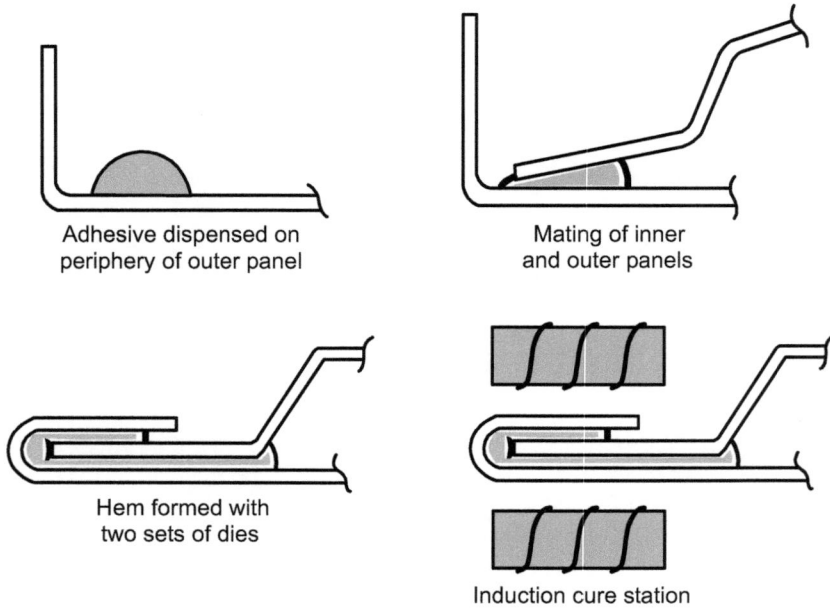

Fig. (11). Induction heating vs standard oven heating.

A two-component induction cure epoxy adhesive 5045, supplied by 3M Corporation, is formulated especially for aluminum hem flange bonding applications as an automotive structural adhesive. Product technical data is shown in Table **7** [25].

Table 7. Typical properties of two-part induction cure epoxy adhesive 5045.

Appearance	
Part A	**Tan**
Part B	**Black**
mixed	**black**
Viscosity, cps@ room temperature	
Part A	31500
Part B	26900

(Table 7) cont.....

Mix ratio of Part A: Part B, by volume	1:2
Open time, minutes	30-60
Induction cure	4-8 seconds @ 79 – 121°C
Over-lap shear strength, MPa	
Room temperature plus oven cure	9.6
Induction plus 30 minutes RT cure	2.07
Induction plus 60 minutes RT cure	2.76
Induction plus 24 RT cure	5.52
Induction plus 10 minutes @130°C oven	
	16.55

LASER CURE EPOXY ADHESIVE TECHNOLOGY

Laser cure epoxy technology is a technology based on the laser radiation heating method to cure epoxy materials. It is a very fast, highly efficient, and indirect heating method with the use of a near-infrared laser beam as the energy source. As shown in Fig. (**12**) [26], the adhesive or/and the adherend adsorb high-energy infrared laser light that can generate heat and cure the adhesive in a few seconds. Laser cure epoxy adhesive is a thermal cure epoxy composition incorporated especially with photothermal conversion material such as carbon black, which can absorb and transfer laser light to heat very efficiently. Recently, significant progress in the development of both laser cure epoxy adhesive and laser cure equipment has been achieved [27 - 30].

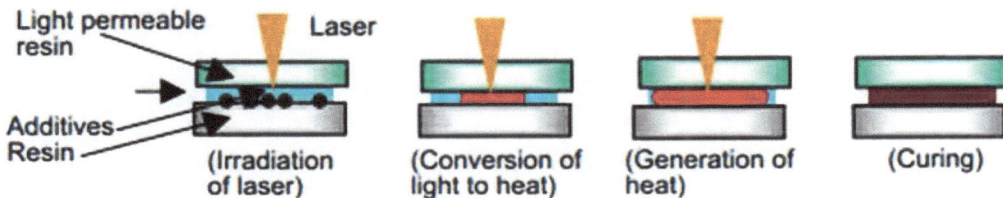

Fig. (12). Laser radiation cure process.

WELD BONDING EPOXY ADHESIVE TECHNOLOGY

Weld bonding epoxy adhesive technology is a hybrid assembly bonding method incorporating conventional welding with epoxy adhesive bonding processes. Weld bonding technology combines instant assembly, strong connection advantages of welding with improved durability resistance, and design flexibility benefits of adhesive bonding. There are mainly two ways of applying adhesive: the flow-through method, by applying adhesive after the weld, and the weld-through method, by applying adhesive before the weld process. Both room temperature and thermal cure epoxy adhesives can be used for weld bonding applications. Weld bonding epoxy adhesive bonding technology has been used in aircraft assembly, automotive production, and vehicle repair and maintenance applications [31 - 34].

CONCLUSION

Most common thermal cure epoxy adhesives can be used to bond metal adherends of iron, steel, copper, and aluminum *via* induction heating. In non-metallic adherend applications, specially designed induction curable epoxy adhesives filled with enough amount of susceptors are used. Iron powder, steel powder, copper powder, and aluminum powder can be used as susceptors. Aluminum powder is most commonly used due to its relatively low density. Induction cure equipment is composed of three units: generator, matching unit, and work coil. Typical applications are magnet bonding for e-motor assembly, hem flange bonding in the automotive industry, and general-purpose bonding applications that require fast production efficiency. Significant progress in the development of both laser cure epoxy adhesive and laser cure equipment has been achieved recently. Weld bonding epoxy adhesive bonding technology has been used in aircraft assembly, automotive production, and vehicle repair and maintenance applications.

REFERENCES

[1]   Petrie, E.M. *Handbook of Adhesives and Sealants*; MicGraw-Hill: New York, **2006**, p. 273.

[2]   Severijns, C.; de Freitas, S.T.; Poulis, J.A. Susceptor-assisted induction curing behaviour of a two component epoxy paste adhesive for aerospace applications. *Int. J. Adhes. Adhes.,* **2017**, *75*, 155-164.
        http://dx.doi.org/10.1016/j.ijadhadh.2017.03.005

[3]   Kowatz, J.; Teutenberg, D.; Meschut, G. Optimization of inductive fast-curing of epoxy adhesive by model-based kinetics. *Int. J. Adhes. Adhes.,* **2023**, *124*, 103392.
        http://dx.doi.org/10.1016/j.ijadhadh.2023.103392

[4]   Eagle, G. High green strength induction curable adhesive. *Japanese Patent 2558281*, **1996**.

[5]   Nakachigi, K.; Toshioka, H. *Japan Patent Application 2007-84767*, **2007**.

[6]   Murata, N. Induction cure adhesive. *Japanese Patent Application H3-45683*, **1991**.

[7]   Eadara, R.; Armbruster, R.F. Induction heat curable epoxy resin systems. *U.S. Patent Application 4992489*, **1991**.

[8]   Wartusch, R.; Herzog, H.; Ortelt, M. Inductively curable adhesive composition. *European Patent Application 2470613*, **2012**.

[9]   Jose, F.; Alexandre, G.; Yann, M.; Damienm, P. Device for assembling parts by bonding by using induction heating. *European Patent 2517532*, **2017**.

[10]   Terry, S.; Raju, R.; Richa, C. Induction heating-cured adhesives. *US Patent Application 20230348656,* **2023**.

[11]   Rudnev, V.; Loveless, D.; Cook, R.L. *Handbook of Induction Heating,* 2nd ed; CRC Press: New York, **2017**.
http://dx.doi.org/10.1201/9781315117485

[12]   Seidensha Electronics Co., Ltd. *Induction Heating Technology,* https://www.sedeco.co.jp/technology/list/ih/ [accessed on Dec 20, 2023].

[13]   Licari, J.J.; Swanson, D.W. *Adhesives Technology for Electric Applications*; William Andrew Publishing: New York, **2005**, pp. 64-73.

[14]   Petrie, E.M. *Handbook of Adhesives and Sealants*; MicGraw-Hill: New York, **2006**, pp. 171-174.

[15]   Nakamura, Y. Modification with inorganic materials. *Epoxy Resins*; Takeuchi, H., Ed.; The Japan Epoxy Technology Society: Tokyo, **2003**, *Vol. 3*, pp. 88-95.

[16]   Samsudin, S.S.; Abdul Majid, M.S.;Ridzuan, M.; Osman, A. Thermal polymer composites of hybrid fillers. *IOP Conf. Series: Mater. Sci. Engineer.,* **2019**, *670*, 012037.

[17]   Shinano Electric Refining Co., Ltd. *Characteristic of SiC.* https://www.shinano-sic.co.jp/en/usage_sic03.html [accessed on Dec 20, 2023].

[18]   Seidensha Electronics Co., Ltd. *Induction Heating Equipment.* https://www.sedeco.co.jp/item/search/list/detail/?pdid=223 [accessed on Dec 20, 2023].

[19]   ThreeBond Holdings Co., Ltd. *Magnet Bonding with an Induction Cure.* https://threebond.com/product-spotlight-magnet-bonding-with-an-induction-cure/ [accessed on Dec 20, 2023].

[20]   Permabond. *Induction Curing Epoxy Adhesives*, https://www.permabond.com/resource-center/induction-curing-epoxy-adhesives/ [accessed on Dec 20, 2023].

[21]   Delo Corporation. *Induction Curing for E-motor/Magnet Bonding.* https://www.youtube.com/watch?v=Z-GQZc-wiDQ [accessed on Dec 20, 2023].

[22]   Nagese Chemtex Corporation. *Permanent Magnet Assembly,* https://nagasechemtex.com/products/industrial/permanent-magnet/ [accessed on Dec 20, 2023].

[23]   Henkel Corporation. *Loctite EA E-220IC,* https://www.gluespec.com/Materials/SpecSheet/33ec1109-fc5b-40eb-9723-f96c2594af32 [accessed on Dec 20, 2023].

[24] Petrie, E.M. *Handbook of Adhesives and Sealants*; MicGraw-Hill: New York, **2006**, p. 275.

[25] 3M Corporation. *Two-part Induction Cure Epoxy Adhesive.* https://multimedia.3m.com/mws/media/494762O/3m-two-part-induction-cure-epoxy-adhesive-5045.pdf [accessed on Dec 20, 2023].

[26] Katsuhiko, S.; Osamu, I.; Kazuo, K. Development of new thermosetting materials and new curing technology. *Denso Technical Review,* **2009**, *14*, 81-87.

[27] Hiramitsu, N.; Komori, T. Realization of high-speed adhesive curing by spot heating and structure consideration of metal adhesive parts. *OMRON Technics,* **2020**, *52*, 1-10.

[28] Nakajima, S.; Tatsuno, Y.; Ito, M.; Hirono, S.; Morizaki, Y.; Kawamoto, J. Curing method of resin composition. Japanese Patent 6331525, **2018**.

[29] Hiramitsu, S.; Komori, Y.; Ito, M.; Hirono, S.; Morizaki, Y.; Kawamoto, J. Production technique of bonding products and bonded products. Japanese Patent 7298348, **2023**.

[30] ADEKA Corporation. *Laser Cure Adhesive System,* https://www.adeka.co.jp/en/development/laboratory/laser [accessed on Dec 20, 2023].

[31] Ghosh, P.K.; Vivek, VIVEK. Welbonding of stainless steel. *ISIJ Int.,* **2003**, *43*(1), 85-94. http://dx.doi.org/10.2355/isijinternational.43.85

[32] Petrie, E.M. Weldbonding – A Hybrid method of assembly. *Met. Finish.,* **2013**, *111*(2), 42-44. http://dx.doi.org/10.1016/S0026-0576(13)70163-6

[33] Santos, I.O.; Zhang, W.; Gonçalves, V.M.; Bay, N.; Martins, P.A.F. Weld bonding of stainless steel. *Int. J. Mach. Tools Manuf.,* **2004**, *44*(14), 1431-1439. http://dx.doi.org/10.1016/j.ijmachtools.2004.06.010

[34] Ely, K.J.; Frech, T.M.; Ritter, G.W. *Method of Weldbonding.* US Patent Application, 2004-0031561, **2004**.

CHAPTER 6

Snap Ambient Cure Epoxy Technology

Abstract: Snap ambient cure epoxy technology is based on two-component room temperature fast cure epoxy adhesive systems, including fast room temperature cure epoxy adhesive, cyanoacrylate hybrid epoxy adhesive, and UV and room temperature cure epoxy adhesive. Their chemistry, cure behavior, key features, and applications are introduced.

Keywords: Cyanoacrylate, Free radical photoinitiator, Mercaptan, Room temperature cure.

FAST ROOM TEMPERATURE CURE EPOXY ADHESIVES

Snap ambient cure epoxy technology is based on two-component room temperature fast curable epoxy adhesive systems, including fast room temperature cure epoxy adhesive, cyanoacrylate hybrid epoxy adhesive, and UV and room temperature cure hybrid epoxy adhesive. Fast room temperature cure epoxy adhesive is primarily composed of epoxy resin as one part and mercaptan-type curing agent with a suitable basic catalyst as the other part. Cyanoacrylate hybrid epoxy adhesive is mainly composed of cyanoacrylate and cationic photoinitiator as one part and epoxy resin as the other part. UV and room temperature cure hybrid epoxy adhesive is mainly composed of epoxy resin, acrylates, and radical photoinitiator as one part and room temperature cure curing agent as the other part.

Mercaptans can cure epoxide very fast in the presence of a basic catalyst at room temperature and thus have been mainly used as curing agents for fast room-temperature epoxy formulations. The cure mechanism of epoxy resin and mercaptan in the existence of an imidazole catalyst is shown in Fig. (**1**) [1]. As a result, the epoxy group reacts with mercaptan almost equivalently to form alcohol and sulfide. Chemical structure, viscosity, CAS No., and SH equivalent of typical commercialized mercaptan-type curing agents are shown in Fig. (**2**).

As can be seen, there are mainly three types of mercaptan-type curing agents: primary thiol group linked to ether as in polymeric mercaptan, primary thiol group linked to ester as shown in TMMP, and secondary thiol group. Reactivity with epoxy resin decreases in the below sequence.

Chunfu Chen

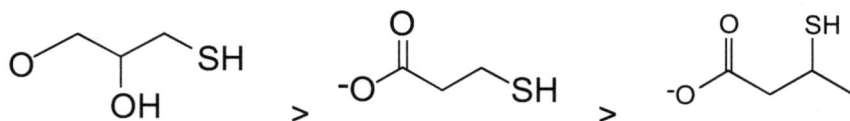

Primary thiol linked to ether   Primary thiol linked to ester   Secondary thiol

a. Initiation

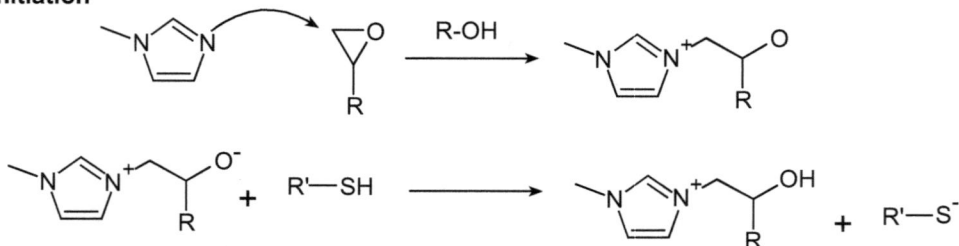

b. Ring-opening of the epoxide

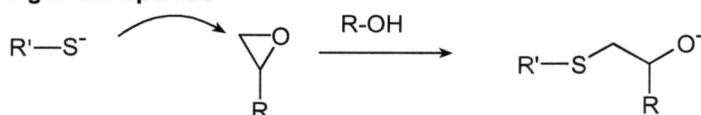

c. Alkoxide/thiol acid-base proton exchange

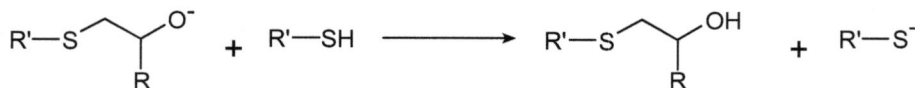

d. Termination / regeneration

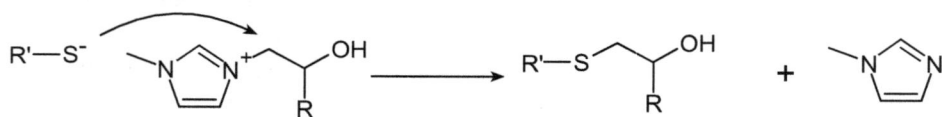

Fig. (1). Polyaddition mechanism of mercaptan with epoxy resin catalyzed by imidazole.

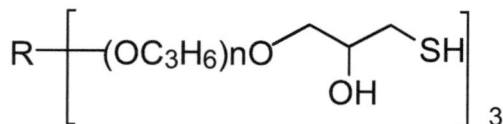

Viscosity: 128°C
Cas No.: 72244-98-5
SH equivalent: 286

Polymercaptan

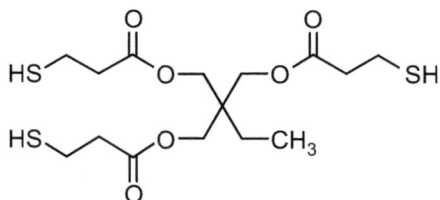

Trimethylolpropane tris(3-mercaptopropionate) (TMMP)

Viscosity: 150 mPa.s/25°C
Cas No.: 33007-83-9
SH equivalent: 141

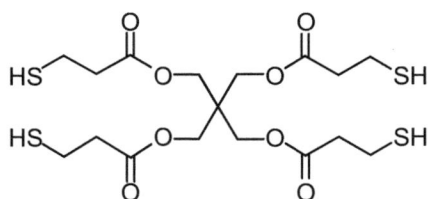

Pentaerythritol tetrakis(3-mercaptopropionate) (PEMP)

Viscosity: 510 mPa.s/25°C
Cas No.: 7275-23-7
SH equivalent: 127

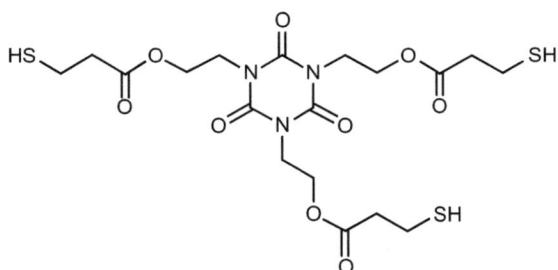

Tris[2-(3-mercaptopropionyloxy)ethyl] Isocyanurate (TEMPIC)

Viscosity: 6000 mPa.s/25°C
Cas No.: 36196-44-8
SH equivalent: 180

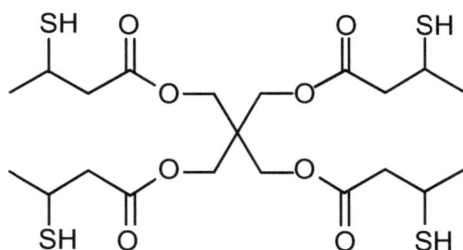

Pentaerythritol tetrakis (3-mercaptobutylate)

Colorless to yellow liquid
Cas No.: 31775-89-0
SH equivalent: 136

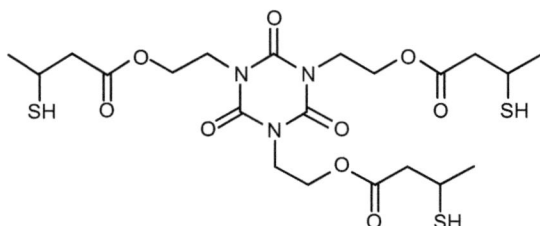

Colorless to yellow liquid
Cas No.: 928339-75-7
SH equivalent: 183-195

1,3,5-Tris [2- (3-mercaptobutanoyloxy) ethyl] -1,3,5-triazine-2,4,6 (1H,3H,5H)
-trione

Fig. (2). Typical mercaptan-type curing agents.

The mercaptan or thiol group reacts with the epoxy group very slowly by itself in the absence of a catalyst at room temperature conditions. Table **1** shows technical data on the commonly used polymeric mercaptan product, Capcure 3-800, supplied by Huntsman Corporation [2]. Its pH value is $3.0 - 5.0$.

Table 1. Technical data on the typical commercial polymeric mercaptan product.

Product Name	CAPCURE® 3-800
Manufacturer	Huntsman Corporation.
Description	A unique polymercaptan epoxy hardener that, when used with a catalyst, provides very rapid cures of epoxy systems, even in thin films and at low temperatures.
CAS No.	72244-98-5
Appearance	Liquid
Odor	Mercaptan
Viscosity, mPas/25°C	10000 - 15000
Density	1.15
pH	$3.0 - 5.0$
SH equivalent	286

(Table 1) cont.....

Chloride content, %	< 0.15
Water content, %	< 0.3

The cure speed of mercaptan with epoxy resin depends very much on the accelerator type and addition amount. Tertiary amines, an imidazole compound, and triphenylphosphine, as shown in Fig. (**3**), are typical accelerators.

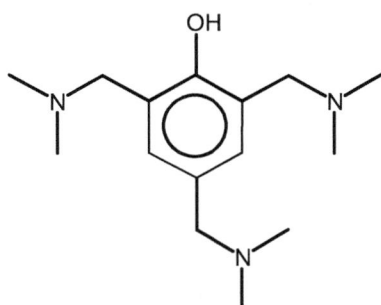

2,4,6-Tris(dimethylaminomethyl)phenol   Triphenyl phosphine

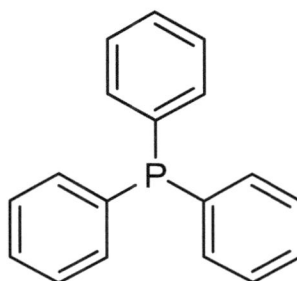

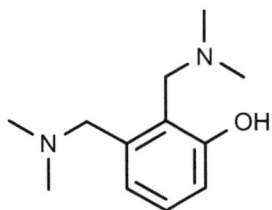

Bis((dimethylamino)methyl)phenol   2-Ethyl-4-methyl imidazole

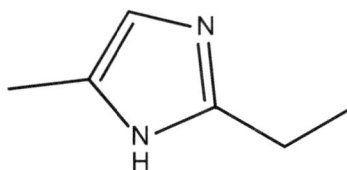

Fig. (3). Chemical structure of typical accelerators for mercaptan curing agent.

By using tertiary amine as the main accelerator, a fast curable mercaptan curing agent, Capcure 3830-31, is developed and supplied by Huntsman Corporation [3]. Table **2** shows the cure behavior, physical properties, and adhesion performance of the fast curable epoxy composition. Gel time at room temperature is quite fast, around 4 minutes. The cured epoxy composition has good physical properties and high adhesion on metal substrates.

Table 2. Fast room temperature cure epoxy composition.

Resin: standard Bisphenol A Epoxy Resin (EEW=190) Hardener: Capcure 3830-81	100 Parts 100 Parts
Gel time, min @ R.T.	4
Tack free time, min @ R.T.	9
Hardness, Shore D, 15 min @ R.T.	83
Tg, °C, cured at R.T. for 7 days	34
Tensile strength, MPa, cured at R.T. for 7 days	54
Elongation, %, cured at R.T. for 7 days	4.5
Lap shear strength, MPa on Al cured at R.T. for 7 days.	19

Table **3** shows a room temperature instant curable two-component epoxy product, Loctite Instant Mix 1 minute epoxy product, supplied by Henkel Corporation [4]. Its gel time at room temperature is only 1 minute, and the product shows satisfactory adhesion performance on various substrates.

Table 3. Loctite Instant Mix 1 minute epoxy product.

Chemical type Resin Hardener	Epoxy resin polymercaptan
Color Resin	Light yellow

(Table 3) cont.....

Hardener	Colorless
Mix ratio, by weight	1:1
Specific Gravity	
Resin	1.16
Hardener	1.12
Gel time, minutes at room temperature	1
Handling time, minutes at room temperature	5 to 10
Full cure time, hours	24
Lap shear strength, MPa on steel, Cured for 24 hours at room temperature.	17.2
Service temperature	
Long term	-23 to 49°C
Short term	-23 to 150°C

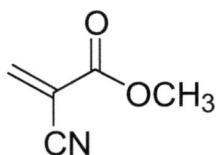

2-Methyl cyanoacrylate

Colorless liquid
CAS No.: 137-05-3
Molar mass: 111

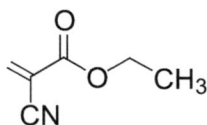

2-Ethyl cyanoacrylate

Colorless to yellow liquid
CAS No.: 7085-85-0
Molar mass: 125

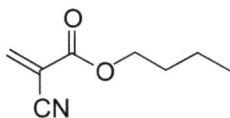

2-Butyl cyanoacrylate

Colorless liquid
CAS No.: 6606-65-1
Molar mass: 153

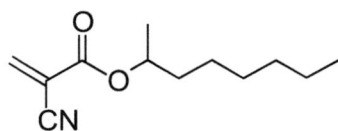

Colorless liquid
CAS No.: 133978-15-1
Molar mass: 209

2-Octyl cyanoacrylate

Fig. (4). Typical alkyl 2-cyanoacrylates.

CYANOACRYLATE HYBRID EPOXY ADHESIVES

Cyanoacrylate adhesive, abbreviated as CA adhesive, is well known as an instant curing adhesive used widely in various bonding and assembly applications nowadays [5-8]. Cyanoacrylate adhesive is mainly based on alkyl 2-cyanoacrylate monomers, whose chemical structure is shown in Fig. (4). As can be seen, there are two strong electronegative cyano (-CN) and carboxyl ester (-COOR) groups connected to the unsaturated C=C double bond, which can thus undergo extremely fast anionic polymerization with weak basic materials such as water and alcohols on adherends to bond the substrates instantly at room temperature. The anionic polymerization mechanism is shown in Fig. (5) [9]. Physical properties and main applications of major alkyl 2-cyanoacrylates are summarized in Table 4. 2-Ethyl cyanoacrylate is the most commonly used cyanoacrylate monomer.

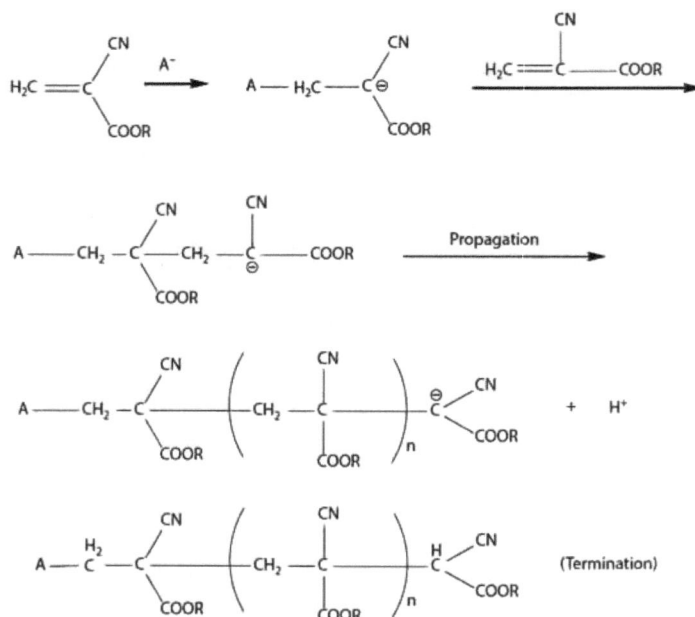

Fig. (5). Anionic polymerization mechanism of alkyl 2-cyanoacrylate.

Table 4. Physical properties and applications of alkyl 2-cyanoacrylate adhesives.

Alkyl	Molecular Weight	Density	Viscosity, mPas/25°C	Surface Tension mN/m	Application
Methyl	111	1.10	2.2	37.41	General purpose
Ethyl	125	1.04	1.86	34.32	
Propyl	139	1.00	1.95	32.8	Rubber bonding
Butyl	153	0.99	2.07	31.11	Medical
Octyl	209	0.93	3.92	29.18	Industry assembly

Due to their high reactivity, alkyl 2-cyanoacrylate monomer and formulated cyanoacrylate adhesives must be well stabilized. Two types of stabilizers are used: anionic inhibitors and free radical inhibitors. Anionic inhibitors include Lewis or protonic acids such as boron trifluoride, maleic acid, and maleic anhydride. Free radical inhibitors are hydroquinone, catechol, hindered phenols, and their derivatives. Stabilizing technology is one key step in the development of cyanoacrylate adhesives.

Cyanoacrylate adhesives are typically one-component systems that do not require an additional premixing process. They are applied as drops directly on the substrate surface and can be set instantly due to small amounts of absorbed moisture, which can initiate its anionic polymerization. Bonding strength and setting time on different substrates by cyanoacrylate adhesive are compared in Table **5**.

Table 5. Bonding efficiency of cyanoacrylate adhesive on different substrates.

Substrates	Bond Strength	Setting Time, Seconds
Aluminum/Aluminum	Intermediate	<= 180
Steel/Steel	Strong	<=180
Steel/Neoprene	Strong	<60

(Table 5) cont.....

Polyethylene/Polyethylene	Intermediate	<=180
PET/PET	Intermediate	<=180
Nylon/Nylon	Intermediate	<60
Glass/Glass	Strong	<60
Glass/Rubber	Strong	<60
Glass/Textile	Strong	<60
Glass/Steel	Strong	<=180
Wood/Wood	Strong	>180
Metal/Leather	Strong	<=180
Metal/Rubber	Strong	<60
Metal/Textile	Strong	<60
Rubber/Rubber	Strong	<60
Rubber/Cardboard	Strong	<60
Porcelain/Porcelain	Strong	<60

A new type of cyanoacrylate hybrid epoxy adhesive was invented and supplied recently [10-14]. This cyanoacrylate hybrid epoxy adhesive is a two-component product comprising a cyanoacrylate component and cationic catalyst as one part and an epoxy component as the other part. Once mixed, the cyanoacrylate component will start to cure fast *via* anionic polymerization, while the epoxy component will cure *via* cationic polymerization initiated by the cationic initiator at room temperature. The cyanoacrylate hybrid epoxy adhesive combines the instant fixture feature of cyanoacrylate well with the high performance of the epoxy composition. The hybrid epoxy adhesive shows much better adhesion reliability performance at both high temperature and high humidity conditions.

Loctite HY 4090 is a two-component, general-purpose cyanoacrylate hybrid epoxy adhesive product supplied by Henkel Corporation [15]. The product is designed to bond a variety of substrates, including metals, most plastics, and rubbers, and can provide good temperature and moisture resistance suitable for use in applications in high-temperature/humidity environments. Typical properties and key features of Loctite HY 4090 are summarized in Table **6**.

Table 6. Loctite HY 4090.

Technology	Cyanoacrylate/Epoxy Hybrid
Components	Two-components – requiring mixing
Part A	Cyanoacrylate
Part B	Epoxy
Mix ratio, by volume Part A: Part B	1:1
Appearance	
Part A	Colorless to straw-colored liquid
Part B	Off-white to light yellow gel
Mixture	Off-white to light yellow gel
Viscosity, mPa.s/25°C	
Part A	4000 – 7000
Part B	25000 - 40000
Fixture time, seconds @25°C	<180
Tg, °C, cured for 168 hours @22°C	88
Coefficient of Thermal Temperature, K^{-1}	
Below Tg	71×10^{-06}
Above Tg	175×10^{-06}
Shore hardness, Durometer D	65 - 69
Tensile strength, MPa	7.1

(Table 6) cont.....

Elongation at break, %	3.6
Lap shear strength, MPa cured for 168 hours@22°C	
Steel (grit blasted)	17
Stainless steel	15
Aluminum	7.6
Aluminum (etched)	13
Zinc dichromate	9.1
ABS	5.2
Phenolic	3.2
Polycarbonate	6.9
Nitrile	0.7
Wood	4.8
Epoxy	9.1
Polyethylene	0.5
Polypropylene	0.6

UV AND ROOM TEMPERATURE CURE EPOXY ADHESIVES

UV cure cationic epoxy and dual cure hybrid epoxy products are usually prepared and supplied in a one-component system for easy use. An additional post-thermal cure process after UV cure is typically required to achieve full cure. In certain situations, however, an additional thermal cure process might be difficult to apply, but the applications will still need fast cure speed and high adhesion performance of the epoxy composition.

Two-component UV and room temperature cure hybrid epoxy systems are developed to meet these uses [16-18]. The hybrid epoxy system is typically composed of epoxy resin, acrylate components, and radical photoinitiator as one part and room temperature curable curing agent component as the other part. Once mixed, the acrylate composition will cure immediately *via* UV light radiation and fix the substrates instantly. The rest of the epoxy composition can achieve full cure in a few days at room temperature, similar to normal room temperature cure two-component epoxy systems, as illustrated in Fig. (**6**).

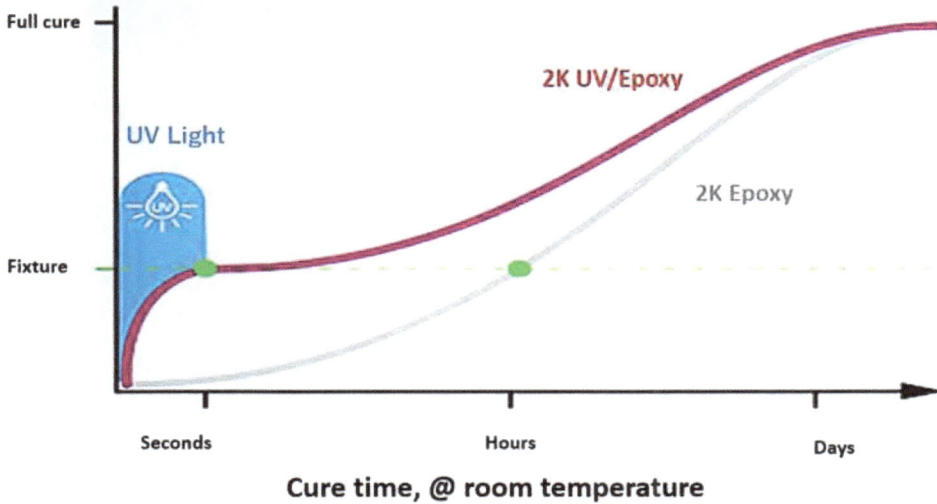

Fig. (6). Fixture time, full cure time comparison of 2K UV/epoxy with 2K epoxy products.

Loctite EA 3336 is a two-component UV and room-temperature curable hybrid epoxy adhesive product supplied by Henkel Corporation [19]. The product can cure fast when exposed to normal UV light. The product has excellent chemical resistance, superior weatherability, and good environmental resistance like normal epoxy products. Typical properties and key features of Loctite EA 3336 are summarized in Table **7**.

Table 7. Typical properties and key features of Loctite EA 3336.

Technology	Acrylate/Epoxy Hybrid
Components	Two-components – requiring mixing
Part A	Acrylate/epoxy

(Table 7) cont.....

Part B	Amine
Mix ratio, by volume Part A: Part B	4:1
Appearance Part A Part B Mixture UV cured	 Colorless to slightly greenish liquid Colorless to slightly yellowish liquid Colorless liquid Colorless to slightly yellowish solid
Viscosity, mPa.s/25°C Part A Part B Mixed	 8000 – 12000 1000 – 3000 14500
Working life, minutes	15
UV fixture time, seconds @100 mW/cm^2 Glass microscope slides used	<5
Tg, °C, cured @ UV 60 mW/cm^2 for 60 seconds, plus 7 days @22°C	55
Shore hardness, Durometer D	79
Tensile strength, MPa	41
Elongation at break, %	6
Lap shear strength, MPa, cured at UV 30 mW/cm^2 for 60 seconds, plus 7 days @22°C Glass to steel	 7.7

(Table 7) cont.....

Glass to G-10 epoxy	5.5
Glass to aluminum	6
Glass to polybutylene terephthalate	4
Polycarbonate to polycarbonate	5.9
Polycarbonate to PVC	2.6

CONCLUSION

Fast room temperature epoxy adhesive, cyanoacrylate hybrid epoxy adhesive, and UV and room temperature cure epoxy adhesive systems are the main commercialized snap ambient cure epoxy technology. Fast room temperature epoxy adhesive is a two-component epoxy composition based on mercaptan-type curing agent with a combination of suitable basic catalysts. Cyanoacrylate hybrid epoxy adhesive is a two-component product comprising a cyanoacrylate component and cationic catalyst as one part and epoxy components as the other part. UV and room temperature cure hybrid epoxy adhesive is typically composed of epoxy resin, acrylate components, and free radical photoinitiator as one part and room temperature curable curing agent component as the other part.

REFERENCES

[1] Fernández-Francos, X.; Konuray, A.O.; Belmonte, A.; De la Flor, S.; Serra, À.; Ramis, X. Sequential curing of off-stoichiometric thiol–epoxy thermosets with a custom-tailored structure. *Polym. Chem.,* **2016**, *7*(12), 2280-2290.
   http://dx.doi.org/10.1039/C6PY00099A

[2] Huntsman Corporation. *Technical Data Sheet of Capcure 3-800 Curing Agent,* https://www.gabrielchem.com/wp-content/uploads/CAPCURE-3-800_US_e.pdf [accessed on Dec 20, 2023].

[3] Huntsman Corporation. *Technical Data Sheet of Capcure 3830-31.* https://www.gabrielchem.com/wp-content/uploads/CAPCURE-3830-81-TDS.pdf [accessed on Dec 20, 2023].

[4] Henkel Corporation. Loctite Epoxy Intstant Mix 1 Minute. https://www.loctiteproducts.com/en/products/build/epoxies/loctite_epoxy_instantmix1 minute.html [accessed on Dec 20, 2023].

[5] Millet, G.H. Cyanoacrylate adhesives.*Structure adhesives*; Hartshorn, S.R., Ed.; Plenum Press: New York, **1986**, pp. 249-307.
   http://dx.doi.org/10.1007/978-1-4684-7781-8_7

[6]   Klemarczyk, P.; Guthrie, J. Advances in anaerobic and cyanoacrylate.*Advances in structural adhesive bonding*; Dillard, D.A., Ed.; Woodhead Publishing: Oxford, **2010**, pp. 96-131.
http://dx.doi.org/10.1533/9781845698058.1.96

[7]   Burns, B. Advances in cyanoacrylate structural adhesives.*Advances in structural adhesive bonding,* 2nd ed; Dillard, D.A., Ed.; Woodhead Publishing: Oxford, **2023**, pp. 137-157.
http://dx.doi.org/10.1016/B978-0-323-91214-3.00017-X

[8]   Coover, H.W.; Dreifus, D.W.; O'Connor, J.T. Cyanoacrylate adhesives.*Handbook of adhesives,* 3rd ed; Skeist, I., Ed.; Chapman & Hall: New York, **1990**, pp. 463-477.
http://dx.doi.org/10.1007/978-1-4613-0671-9_27

[9]   Raja, P.R. Cyanoacrylate adhesives: a critical review. *Rev. Reviews of Adhesion and Adhesives,* **2016**, *4*(4), 398-416.
http://dx.doi.org/10.7569/RAA.2016.097315

[10]   Hersee, R.M.; Burns, B.; Barnes, R.B.; Tully, R.P.; Guthrie, J. *Two-part Cyanoacrylate/Cationically Curable Adhesive System.* U.S. Patent 8742048, **2014**.

[11]   Hersee, R.M.; Burns, B.; Barnes, R.B.; Tully, R.P.; Guthrie, J. *Two-part Cyanoacrylate/Cationically Curable Adhesive Syste*m. European Patent 2616520, **2018**.

[12]   Singh, S.; Jayesh, J.; Tale, N.; Trivedi, K. A two-part cyanoacrylate curable adhesive system. *U.K. Patent 2576792,* **2022**.

[13]   Lavoie, N. Innovations in hybrid structural instant adhesive technologies. *Plastics Engineering -Connecticut,* **2016**, *72*(2), 28-35.

[14]   Goss, B. Hybrid Adhesives - *The Best of all Worlds.* EURADH 2016 - Adhesion '16, Glasgow, UK **2016**.

[15]   Henkel Corporation. *Loctite HY 4090 Universal Structural Bonder.* https://www.henkel-adhesives.com/us/en/product/structural-adhesives/loctite_hy_40900.html [accessed on Dec 20, 2023].

[16]   DELO Corporation. Fixing Tow-component Epoxy Resins Under UV Light https://www.delo-adhesives.com/hybrid-chemistry [accessed on Dec 20, 2023].

[17]   Nativi, L.A.; Kropp, P.L. *Two Component Curable Epoxy Resin Composition having a Long Pot Life.* European Patent 0245559, **1991**.

[18]   Zhou, J.; Wang, X.; He, X.; Zhao, Y. *UV curable Two-component Epoxy Adhesive.* Chinese Patent Application 111826109, **2020**.

[19]   Henkel   Corporation.   *Loctite   EA   3336.*   https://www.henkel-adhesives.com/fr/en/product/light-curing-adhesives/loctite_ea_3336.html [accessed on Dec 20, 2023].

SUBJECT INDEX

A

Absorbency property 91, 125
Absorption 79, 80, 117, 118, 120
   spectra of acylphosphine oxide
      photoinitiator TPO 118
   spectra of benzophenone 120
Accelerator 31, 37, 40, 41, 57, 134, 135, 138,
      139, 140, 144, 147, 150, 160
  internal phenolic 31
ACF 157, 159
   composition 157, 159
   technology 157
Acid(s) 2, 3, 16, 75, 79, 81, 89, 90, 98, 100,
      111, 112, 191
   acrylic 111, 112
   bronsted 75, 79, 98
   carboxylic 16
   maleic 191
   protonic 191
ACP technology 157
Acrylate(s) 59, 61, 62, 104, 105, 106, 108,
      110, 111, 120, 123, 125, 126, 130, 183,
      194, 197
   aliphatic urethane 111
   aromatic urethane 110
   components 59, 104, 123, 194, 197
   monomer curing 125
   multi-functional 105, 106
   oligomer 104, 110, 123, 130
Acrylate monomers 104, 105, 108, 109, 110,
      112, 120, 122, 124, 128, 130
   multi-functional 108
Acrylic polymer network 120
Active amine hydrogen 24
Active hydrogen 22, 23, 24, 42, 138, 145, 146,
     147
  polyaddition mechanism of 146
Adduct reaction of polyamine 29
Adeka resin products 13
Adherend(s) 92, 123, 125, 151, 165, 166, 179,
     190

bonding metal 166
bonding non-metallic 166
   substrate 92, 151
Adhesion 45, 58, 89, 94
  behavior 94
Adhesive 1, 50, 52, 53, 54, 57, 58, 96, 152,
     192, 195
  film 57, 58
  product 50, 52, 53, 54, 96, 152, 192, 195
  technology developments 1
Aerospace assembly applications 55
Air atmosphere 130
Aircraft assembly automobile production 61
Aliphatic polyamine 23, 24, 51, 146, 147
Aluminum 44, 51, 53, 54, 165, 166, 169, 170,
     171, 172, 177, 178, 180
  metallic 169
  powder 166, 180
Amine 1, 2, 3, 10, 23, 24, 26, 27, 29, 31, 42,
     51, 55, 59, 138, 149, 196
  adducts 29, 31, 51
  curing agent 24, 59
  cycloaliphatic 23, 26, 27, 55
Analytical methods 94
Ancamide products 30
Anhydride 1, 16, 17, 34, 35, 36, 37, 42, 144,
     146, 191
  maleic 191
  methyl himic 35
Anionic polymerization 22, 37, 61, 138, 190,
     191, 192
  mechanism 22, 190
  of cyanoacrylate 61
  tertiary amine-catalyzed 37
Anisotropic conductive 134, 157, 160
   adhesives (ACAs) 157
   film (ACF) 134, 157, 160
   paste (ACP) 157
Applications 1, 6, 21, 34, 96, 127, 134, 138,
     157, 166, 176, 177, 180, 183, 190, 191,
     193, 194
  adhesive 34

U

V

W

www.ingramcontent.com/pod-product-compliance
Lightning Source LLC
Chambersburg PA
CBHW050845220326
41598CB00006B/438